# Wireless Network Pricing

# Synthesis Lectures on Communication Networks

Editor
**Jean Walrand,** *University of California, Berkeley*

Synthesis Lectures on Communication Networks is an ongoing series of 50- to 100-page publications on topics on the design, implementation, and management of communication networks. Each lecture is a self-contained presentation of one topic by a leading expert. The topics range from algorithms to hardware implementations and cover a broad spectrum of issues from security to multiple-access protocols. The series addresses technologies from sensor networks to reconfigurable optical networks.
The series is designed to:

- Provide the best available presentations of important aspects of communication networks.

- Help engineers and advanced students keep up with recent developments in a rapidly evolving technology.

- Facilitate the development of courses in this field

Wireless Network Pricing
Jianwei Huang and Lin Gao
2013

Performance Modeling, Stochastic Networks, and Statistical Multiplexing, second edition
Ravi R. Mazumdar
2013

Packets with Deadlines: A Framework for Real-Time Wireless Networks
I-Hong Hou and P.R. Kumar
2013

Energy-Efficient Scheduling under Delay Constraints for Wireless Networks
Randall Berry, Eytan Modiano, and Murtaza Zafer
2012

NS Simulator for Beginners
Eitan Altman and Tania Jiménez
2012

Network Games: Theory, Models, and Dynamics
Ishai Menache and Asuman Ozdaglar
2011

An Introduction to Models of Online Peer-to-Peer Social Networking
George Kesidis
2010

Stochastic Network Optimization with Application to Communication and Queueing
Systems
Michael J. Neely
2010

Scheduling and Congestion Control for Wireless and Processing Networks
Libin Jiang and Jean Walrand
2010

Performance Modeling of Communication Networks with Markov Chains
Jeonghoon Mo
2010

Communication Networks: A Concise Introduction
Jean Walrand and Shyam Parekh
2010

Path Problems in Networks
John S. Baras and George Theodorakopoulos
2010

Performance Modeling, Loss Networks, and Statistical Multiplexing
Ravi R. Mazumdar
2009

Network Simulation
Richard M. Fujimoto, Kalyan S. Perumalla, and George F. Riley
2006

Wireless Network Pricing

Jianwei Huang and Lin Gao

ISBN: 978-3-031-79262-5    paperback
ISBN: 978-3-031-79263-2    ebook

DOI 10.1007/978-3-031-79263-2

A Publication in the Springer series
*SYNTHESIS LECTURES ON COMMUNICATION NETWORKS*

Lecture #13
Series Editor: Jean Walrand, *University of California, Berkeley*
Series ISSN
Synthesis Lectures on Communication Networks
Print 1935-4185    Electronic 1935-4193

# Wireless Network Pricing

Jianwei Huang and Lin Gao
The Chinese University of Hong Kong

*SYNTHESIS LECTURES ON COMMUNICATION NETWORKS #13*

# ABSTRACT

Today's wireless communications and networking practices are tightly coupled with economic considerations, to the extent that it is almost impossible to make a sound technology choice without understanding the corresponding economic implications. This book aims at providing a foundational introduction on how microeconomics, and pricing theory in particular, can help us to understand and build better wireless networks. The book can be used as lecture notes for a course in the field of network economics, or a reference book for wireless engineers and applied economists to understand how pricing mechanisms influence the fast growing modern wireless industry.

This book first covers the basics of wireless communication technologies and microeconomics, before going in-depth about several pricing models and their wireless applications. The pricing models include social optimal pricing, monopoly pricing, price differentiation, oligopoly pricing, and network externalities, supported by introductory discussions of convex optimization and game theory. The wireless applications include wireless video streaming, service provider competitions, cellular usage-based pricing, network partial price differentiation, wireless spectrum leasing, distributed power control, and cellular technology upgrade. More information related to the book (including references, slides, and videos) can be found at `http://ncel.ie.cuhk.edu.hk/content/wireless-network-pricing`.

# KEYWORDS

network economics, network pricing, game theory, monopoly, oligopoly, price discriminations, market competition, network externalities, network effect, network optimization, distributed algorithms, wireless communications, wireless networks, resource allocation, power control, cognitive radio, dynamic spectrum management, femtocell economics, network upgrade

*To my parents and brother, for teaching me the meaning of life*
*To Biying and William, for your love*

*– Jianwei Huang*

*To my parents, my wife, and my daughter*

*– Lin Gao*

# Contents

# Preface

Today's wireless communications and networking practices are tightly coupled with economic considerations, to the extent that it is almost impossible to make a sound technology choice without understanding its economic implications. In this book, we will focus on how pricing theory will help us to understand and build better wireless networks.

We start in Chapter 1 by discussing the motivation for us to write (and for you to read) this book. In particular, we show that economic mechanisms are becoming indispensable parts of the wireless network planning and operating, mainly due to the inherent conflict between the limited wireless resources and the fast growing wireless demands.

Chapters 2 and 3 provide some basic knowledge of the wireless technologies and microeconomics, paving ways for the more exciting and advanced discussions later on.

From Chapter 4 to Chapter 7, we introduce several key aspects of wireless network pricing. In each chapter, we first introduce the related theoretical background, then give two applications in wireless communications and networking to illustrate the theory, finally provide several exercises to test the readers' understanding of the key concepts. The applications in these chapters are for illustration purposes only, and the choices are biased based on our own research interests. More specifically, Chapter 4 focuses on social optimal pricing. Chapter 5 looks at the issue of monopoly, where a single service provider dominates the market and wants to maximize its profit, either through uniform pricing or price differentiation. Chapter 6 concerns the price competition among multiple service providers. Chapter 7 talks about the issue of network externalities. Finally, in Chapter 8, we come back to the larger topic of wireless network economics, and discuss the connections between pricing and several other economic mechanisms such as auction, contract, and bargaining.

This book is intended as a reference for graduate students and senior undergraduate students when taking a course on network economics, for researchers who are interested in the fundamental issues and solution tools in this area, for wireless engineers who need to understand economic principles to better design and control networks, and for applied economists who are curious about how microeconomics is making an impact in the rapid growing wireless industry. Readers can find references, papers, slides, and videos related to this book at the companion website: `http://ncel.ie.cuhk.edu.hk/content/wireless-network-pricing`.

Jianwei Huang and Lin Gao
May 2013

# Acknowledgments

We thank the series editor, Prof. Jean Walrand, for encouraging us to start this book project during JWH's 2010 summer visit in UC Berkeley. We thank co-authors of the publications that form the basis of wireless applications introduced in this book: Mung Chiang, Lingjie Duan, Vojislav Gajic, Aggelos K. Katsaggelos, Shuqin Helen Li, Shuo-Yen Robert Li, Zhu Li, and Bixio Rimoldi. We thank members of the Network Communications and Economics Lab (NCEL) in the department of Information Engineering at the Chinese University of Hong Kong. Many NCELers, including Man Hon Cheung, Quansheng Guan, Changkun Jiang, Yuan Luo, Qian Ma, Richard Southwell, Hao Wang, Haran Yu, and Junlin Yu, have contributed to the book by providing critical and constructive comments. We thank Tianyi Hu, for helping draw many of the figures in the theory part of this book. JWH would like to thank his Ph.D. advisors, Randall Berry and Michael Honig, for bringing him into the fascinating world of network economics. This work is supported by the General Research Funds (Project Number CUHK 412710 and CUHK 412511) established under the University Grant Committee of the Hong Kong Special Administrative Region, China.

Jianwei Huang and Lin Gao
Shatin, Hong Kong
May 2013

# CHAPTER 1

# Introduction

## 1.1 WHY THIS BOOK?

Today's wireless communications and networking practices are tightly coupled with economic considerations, to the extent that it is almost impossible to make a sound technology choice without understanding its economic implications. This simple fact motivates us to take a close and systematic look at how economics interacts with wireless technologies. In this chapter, we will outline the big picture of wireless network economics, centered around the following two questions:

- Why should we care about economics in wireless networks?

- What are the unique challenges of wireless network economics?

We want to point out that wireless network economics is a vast topic that is difficult to cover in fewer than 200 pages, especially if we want to provide concrete examples with some analytical details. Therefore, in the rest of the book, we choose to focus on one key aspect of wireless network economic—wireless network pricing—to give readers a partial but hopefully more focused and in-depth view of the challenging economic issues of the wireless networks.

## 1.2 THE WIRELESS REALITY

Let us first imagine a "wireless utopia," where the wireless spectrum is unlimited, the wireless technologies can provide a communication speed comparable to wireline networks, heterogeneous wireless technologies co-exist perfectly without mutual interferences, wireless users have reasonable demands that can always be satisfied without overburdening the network, and wireless service providers aim to maximize the social welfare instead of their own profits. In this perfect world, every user can enjoy seamless and high speed wireless connections whenever and wherever they want, and there is no reason to worry about economic issues.

However, the reality of wireless networks is (almost) exactly the opposite. The wireless spectrum is very limited and overly crowded, the communication speed of the latest wireless technologies is nowhere close to that of wireline networks (except for some very short distance wireless communications), heterogeneous wireless networks often exist with little or no coordinations, heavy mutual interferences between networks and devices are the norm rather than the exceptions, the exploding growth of wireless data traffic is far beyond the growth of wireless capacity, and the wireless service providers often care more about profits than social welfare. Some of the above issues can be alleviated by the advance of wireless technologies; many others can only be addressed with a combination of technology advances, economic innovations, and policy reforms.

Next we will illustrate several of the above issues in a bit more detail, and outline how economics can help to improve the overall performance of the wireless networks and satisfaction levels for both users and service providers.

## 1.3   TENSION BETWEEN SUPPLY AND DEMAND

One key reason for studying wireless network economics is to resolve the tension between limited wireless resource supplies and the fast growing wireless demands.

The radio spectrum is limited, and only a fraction of it (mostly the lower frequency part) is useful for wireless communications over reasonable ranges. Because of the limited availability of wireless spectrum, it has been a tightly controlled resource worldwide since the early part of the 20th century. The traditional way of regulating the spectrum is the static licensing approach, which assigns each wireless application a particular piece of spectrum at each particular location. Currently, almost all spectrum licenses belong to government identities and commercial operators. This can be clearly shown in the frequency allocation map of any country or region.

However, new wireless technologies and services are emerging rapidly. This means that every new wireless commercial service, from satellite broadcasting to wireless local-area network, has to compete for licenses with numerous existing sources, creating a state of spectrum drought [1].

A key challenge for government regulators is how to allocate these ever decreasing and precious spectrum resources wisely to achieve the maximum benefits for society. Among many possible solutions, the spectrum auction has been advocated and successfully implemented in many countries. This will help to allocate the spectrum to service providers who value the resources most, as these providers are typically the ones who have the best technologies and thus the capability to provide the maximum benefits to the customers.

A more revolutionary approach is to enable unlicensed wireless users to opportunistically share the spectrum with licensed users through dynamic spectrum management. This is motivated by the fact that many licensed spectrum bands are not efficiently utilized [2]. For example, the Federal Communications Commission (FCC) in the US has recently decided to open up the TV spectrum for unlicensed use, as long as the licensed users' communications are protected. Microsoft has already built a testbed over its Redmond campus to demonstrate the practicality of such sharing [3].

Note that there are two economic issues under this dynamic spectrum management regime. First, the regulators need to provide enough economic incentives for the license holders to open up spectrum for sharing, otherwise complicated legal issues might arise. The law suit between FCC and National Association of Broadcasters in 2009 is a good example [4]. Second, it remains an open question as to what kind of services and commercial business models can succeed in the newly open spectrum bands, considering the potentially unregulated interferences among multiple unlicensed service providers [5].

The other perspective of the limited wireless spectrum is the tension between the low and often unreliable data rates provided by today's wireless technologies and the fast growing needs of wireless users. One may argue that the Wi-Fi technology (e.g., the IEEE 802.11 family) can already provide a speed of hundreds of Mbps, which is good enough even for high definition video streaming. However, the Wi-Fi technology has a very limited coverage (e.g., from 20 to 200 meters for indoor communications), and thus cannot provide a ubiquitous wireless access experience. The cellular network still remains as the only wireless technology that has the potential to provide seamless access and mobility solutions. Today's 4G cellular networks can provide a theoretical peak download speed of 100 Mbps, although the actual speed can be less than 10% of the theoretical one. The speed per user will be even less when many users share resources of a same base station, which is often the case in practice. On the other hand, thanks to the introduction of sophisticated smartphones and tablets, users have significantly higher needs to enjoy high quality and highly interactive content on-the-go. Consider, for example, the very popular video streaming application of Netflix, which has been available on the iPad platform since 2010. To stream a high quality video, Netflix recommends a data rate of at least 5Mbps. An always smooth playback requires the data rate to be much higher. Applications like these make the current cellular network very stressful. It is widely known that AT&T networks in big US cities such as New York City and San Francisco often have experienced heavy congestion and low achievable data rates during the past several years, ever since AT&T introduced iPhone on their networks from 2007. During the Christmas season of 2009, AT&T even tentatively stopped selling iPhones in New York City, and many suspected that it was due to AT&T's fear of not being able to support the fast growing population of new iPhone users.

Due to the limited spectrum and the constraints of today's wireless cellular technologies, it is impossible to over-provision the wireless network as we did for fiber-based wireline networks. In other words, technology advance alone is not enough to resolve the tension between the supply and demand in the wireless market even in the long run. It is thus very important to use economics to guide the operation of the market.

## 1.4    COUPLING BETWEEN ECONOMICS AND WIRELESS TECHNOLOGIES

The economics of wireless networks can be quite different from economics of other industries, mainly due to the unique characteristics of the wireless technologies and applications.

From the wireless technology side, there are many choices today in the market, and each has its unique strength and weaknesses. For example, Wi-Fi technology can provide high data rates within a short distance, and the cellular technology provides much better coverage with a much lower data rate. The economic models for these two technologies are thus very different. In practice, commercial Wi-Fi providers often charge users based on connection time lengths, while cellular service providers often charge users based on their actual data usage.

In terms of wireless applications, each application (and the user behind the application) has a unique Quality of Service requirement, resource implication on the networks, and sensitivity to price. For example, a video streaming application requires a wireless connection that supports a high data rate and stringent delay requirements. It is possible to charge a high price for such an inelastic application. However, providing a videos streaming application with a data rate higher than needed will not be useful. A file transfer application can adapt to different transmission speeds, but requires a very low bit error rate to ensure correct decoding. Such elastic application will be very sensitive to price, and can be arranged to be delivered when the network is not congested and the delivery cost per bit is low.

The key challenge of wireless network economics is to properly match the wireless technologies with the wireless applications via the most proper economical mechanisms. We also want to emphasize that the choices of wireless technologies and applications are not static. Which technology and application will dominate the market at what time will also heavily depend on the economic implications. For example, although the 4G cellular technology has been available to many operators globally, only a small number of operators have upgraded to 4G networks already. The factors to be considered include how the upgrade costs evolve over time, how fast the users will accept the 4G technology, what types of applications will emerge and fully take advantage of the new technology, how the market competition will affect the pricing strategies, and how the network effect will affect the value of the new service [7]. Thus, it is critical to understand the impact of economics on the evolution of wireless network technologies and applications.

## 1.5  EFFECT OF MARKET DEREGULATION

The deregulation of telecommunication markets in many countries has made the study of wireless network economics more important than ever. In the past, very often there was only one major wireless service provider enjoying the monopoly status in a particular local (or national) market. Examples including AT&T in the US, China Mobile in China, and Telcel in Mexico. However, the recent telecommunication deregulation leads to several major players in a single market. Examples include AT&T, Verizon, T-Mobile, and Sprint in the US, as well as China Mobile, China Unicom, and China Telecom in China. As a wireless service provider is ultimately a profit-maximizing entity, it needs to optimize the technology choices and pricing mechanisms under the intense market competition.

Industry deregulation also brings more choices to the wireless consumers. For example, a user may freely compare and choose services from different service providers based on the service quality and cost. A user may even use different service providers for different types of services, such as using both cellular service and Wi-Fi service through the same cell phone. A service provider may no longer have complete control of its subscribers. All these bring interesting and sometimes new economic questions that are not present in other industries.

## 1.6  WE ARE TALKING ABOUT WIRELESS

One may argue that researchers have studied Internet economics for more than a decade, and the lessons and results learned there can be carried over to the wireless industry. However, wireless network economics is significantly different from Internet economics in several ways.

First, the characterization of network resources in wireless networks is more difficult than in wireline networks. Although wireless spectrum can be measured in hertz, the network resource corresponding to each hertz of spectrum is not easy to characterize. For example, the wireless data rate is often highly stochastic over time due to shadowing, fading, and mobility. Furthermore, the wireless resource is spatially heterogeneous, and the same spectrum may be concurrently used by multiple users who are physically far apart without affecting each other. Finally, the wireless data rates are affected by mutual interferences. Although there are many analytical models characterizing the interference relationships, they can be either clean yet imprecise (e.g., the protocol interference model) or precise yet complicated (e.g., the signal-to-interference-plus-noise ratio [SINR] model). There does not yet exist a model that is precise and analytically tractable for all practical wireless networks.

Second, the characterization of end users can be more complicated in wireless networks. A wireless user may have many different attributes, such as utility function (determined by the application type), total energy constraint and energy efficiency (determined by battery technology and charging levels), and channel conditions (determined by node locations and mobility). Also, users' performances are often tightly coupled due to mutual interferences.

Third, the interactions between wireless users heavily depend on the specific choice of wireless technology. In random medium access protocols such as the slotted Aloha, users are coupled through their channel access probabilities. In Code Division Multiple Access (CDMA) network, users are coupled through mutual interferences. When we consider a spectrum overlay in cognitive radio networks, unlicensed users cannot transmit simultaneously with the licensed users in the same channel at the same location. In a spectrum underlay cognitive radio network, unlicensed users are allowed to transmit simultaneously with the licensed users, as long as the total unlicensed interference generated at a particular licensed receiver is below an interference threshold. Different interactions and couplings between users lead to different types of markets and economic mechanisms.

Fourth, the coupling between technology, policy, and economics is different in wireless networks. We can use cognitive radio as an example to illustrate this point. Cognitive radio technology enables more flexible radio transmitters and receivers, and makes it feasible for wireless devices to sense and opportunistically utilize the spectrum holes. However, how and when cognitive radio technology should be used heavily depends on the type of spectrum band, which determines the types of licensed users and how they value the pros and cons of the new technology. Regulators in some countries are also more conservative than others in approving the new technology and changing the existing licensing practice. In fact, many wireless technologies can only work under the proper policy framework together with the right economic mechanisms that incentivize all parties involved.

## 1.7    SUGGESTIONS FOR READING THIS BOOK

This book includes a total of eight chapters. Figure 1.1 illustrates the interdependency relationship among various chapters.

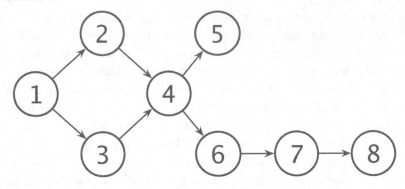

**Figure 1.1:** The interdependency relationship among eight chapters in this book.

After finishing Chapter 1, one can choose to read both or just one of Chapters 2 and 3 depending on his/her technical background. In Chapter 2, we introduce some basic terminologies of wireless communications and networking. This is mainly intended for readers with almost no background in wireless communications. In Chapter 3, we introduce the basics of microeconomics. This will prepare readers with little economics background a good foundation for understanding the more advanced economic models in later chapters.

From Chapter 4 to Chapter 7, we introduce the key aspects of wireless network pricing one by one. All chapters follow a similar structure. We first introduce the related theoretical background, then give two applications in wireless communications and networking to illustrate the theory, and provide several exercises to test the readers' understanding of the key concepts.

We recommend everyone to carefully read through Chapter 4, which introduces the theory of convex optimization, dual-based algorithms, and social optimal pricing, and hence will be useful in understanding all later chapters. Following Chapter 4, one can continue to read Chapter 5, which discusses the issue of monopoly pricing, where a single service provider dominates the market and prices the resources to maximize its profit. One may also skip Chapter 5 and directly read Chapter 6, which concerns the price competition among multiple service providers. Chapter 7 further discusses about the issue of network externalities, sometimes with competing service providers. Finally, in Chapter 8, we come back to the larger topic of wireless network economics, and discuss the connections between pricing and several other economic mechanisms such as auction, contract, and bargaining. Most of these mechanisms are related to games theory illustrated in Chapters 6 and 7.

We also invite readers to visit the companion website of the book: http://ncel.ie.cuhk. edu.hk/content/wireless-network-pricing, where one can find references, papers, slides, and videos related to this book.

CHAPTER 2

# Wireless Communications Basics

In this chapter, we will provide a brief introduction to wireless communications and networking technologies, including the radio propagation characteristics and channel models for wireless communications, the wireless access and networking technologies, and the radio resource management technologies in wireless networks. The intention is to cover the most basic features of wireless communications and networking, in order to reduce readers' need to consult other textbooks to understand the applications in the later chapters of this book. Most discussions in this chapter are based on the materials in [8, 9, 10, 11, 12, 13, 14, 15].

## 2.1 WIRELESS COMMUNICATIONS

Wireless communication refers to the information transfer between two or more points that are not connected by an electrical conductor. The most common wireless communication is done through transmission of signals through free space by electromagnetic radiation of frequencies in the range of around 30 kHz to 300 GHz. These *radio waves* are transmitted and received through *antennas*, which convert electric currents into radio waves, and vice versa. Typical examples of wireless communication systems include mobile cellular networks, wireless LAN (WiFi) systems, broadcast and television systems, global positioning system (GPS), etc.

### 2.1.1 RADIO PROPAGATION

Radio propagation refers to the transmission of radio waves from one point to another. As a form of electromagnetic radiation, radio waves travel through space either straightly and directly (called line-of-sight propagation), or in a path affected by reflection, refraction, diffraction, absorption, polarization, and scattering. Here we summarize the most common radio propagation behaviors as follows:

- *Line-of-sight (LOS) propagation* usually refers to the direct propagation of radio waves between two points that are visible to each other. Examples of LOS propagation include the propagation between a satellite and a ground antenna, and the reception of television signals from a local TV transmitter. The key feature of LOS propagation is that the power density of a radio wave is proportional to the inverse of the square of the transmission distance.

- *Reflection* arises when a radio wave hits the interface between two dissimilar media, so that all of or at least part of the wave returns into the medium from which it originated. With the effect of reflection, a radio wave attenuates by a factor, depending on the frequency, the angle of incidence, and the nature of the medium (e.g., material properties, thickness, homogeneity, etc.). Research shows that the reflection effect often dominates the propagation of radio in indoor scenarios.

- *Diffraction* usually refers to the propagation of radio waves bending around corners or sharp edges. Due to the effect of diffraction, for example, we can hear sound from sources that are out of sight around a corner. Radio waves can bend over hills and around buildings, and therefore get to the shadowed regions where the LOS path is not available. The diffracted radio wave is usually much weaker than that experienced via reflection or direct transmission. Thus, the diffraction propagation is more significant in outdoor scenarios, but is less significant in indoor scenarios where the diffracted signal is much weaker than the reflected signal.

- *Scattering* is a physical effect of radio waves when hitting irregular objects (e.g., walls with rough surfaces). Instead of proceeding in a straight trajectory, the radio wave is redistributed in all directions due to the reflection or refraction by the microscopic textures in the object. This angular redistribution is referred to as the scattering effect. Scattering waves in many directions usually results in reduced power levels, especially far from the scatterer.

These effects, together with many other effects such as refraction, absorption, and polarization, generally co-exist (but with different significances) in radio propagation. Which one dominates the propagation process essentially depends on the particular scenario (e.g., indoor or outdoor). However, it is impossible to perfectly characterize or fully predict the propagation of a radio wave, due to the rapid fluctuation of radio propagation.

In general, radio propagation can be roughly characterized in the large and small scale. Specifically, the large-scale propagation model mainly characterizes the mean attenuation of radio waves over large travel distances; while the small-scale propagation model mainly characterizes the fast fluctuations of radio waves over very short travel distances (e.g., a few wavelengths) or short time durations (e.g., a few milliseconds). More precisely, we can often characterize radio propagation by the following three nearly independent factors: (i) distance-based path loss, (ii) slow log-normal shadowing, and (iii) fast multi-path fading. The first two factors are known as the large-scale propagation, and the last factor is shown as the small-scale propagation. The model of such large- and small-scale propagations is usually called the *channel model*. Having an accurate channel model is extremely important for analyzing and designing a wireless communication system.

## 2.1.2  CHANNEL MODEL

When radio wave propagates, its power density diminishes gradually. In addition, noise may pollute the desired signal, generating the so-called *interference*. The noise or interference may come from natural sources, as well as artificial sources such as the signals of other transmitters. In general, the power profile of the received signal can be obtained by multiplying the power profile of the transmitted signal with the impulse response of the channel. That is, given the transmitted signal $x$, after propagation through a channel $h$, we have the following received signal $y$:

$$y = h \cdot x + \epsilon, \tag{2.1}$$

where $h$ is the channel impulse response, and $\epsilon$ is the noise. Note that the noise $\epsilon$ is usually modeled as a random variable following the normal distribution (also called Gaussian distribution). Thus, it is also referred to as the Gaussian noise. Besides, the noise $\epsilon$ at different time instances is usually assumed to be identically distributed and statistically independent. In this case, the noise is usually called the *white Gaussian noise*.

The purpose of channel modeling is to partially or fully characterize the radio propagation, or in other words, to accurately predict the channel impulse response. According to the features of radio propagation, the channel impulse response mainly depends on the following three key components: *path loss*, *shadowing*, and *multi-path propagation*.

- *Path Loss* is a fundamental characteristic of radio wave propagation in order to predict the expected received power. Traditionally, path loss is examined by using the Friis transmission formula, which provides a means for predicting this received power:

$$|h| = \frac{P_r}{P_t} = A \cdot \frac{\lambda^2}{d^2}, \tag{2.2}$$

  where $\lambda$ is the wavelength, $d$ is the transmission distance, and $A$ is a constant (which is independent of propagation, but is related to the parameters such as antenna gains, antenna losses, filter losses, etc.). The path loss is directly obtained from (2.2) by changing the power unit into dB:

$$\overline{PL}(d) = -10\log(|h|) = 20\log(d) - 20\log(\lambda) - 10\log(A). \tag{2.3}$$

  By (2.2) or (2.3), we can see that radio waves have different propagation losses with different frequencies. For example, radio waves at the lower frequency band (i.e., with a larger wavelength $\lambda$) have a lower propagation loss, and thus are more suitable for long range communications; while radio waves at the higher frequency band (i.e., with a smaller wavelength $\lambda$) are more suitable for short-range but high-speed wireless communications. We can further see that the power falls off in proportion to the square of the distance $d$. In practice, however, the power may fall off more quickly, typically following the 3rd or 4th power of distance $d$.

- *Shadowing* is another important characteristic of radio wave propagation that characterizes the deviation of the received power about the average received power. It usually occurs when a large obstruction (such as a hill or large building) obscures the main propagation path between the transmitter and the receiver. In this case, the received power may deviate from the distance-dependent power given in (2.3), due to the effects of reflection, diffraction, scattering, etc. This is usually referred to as the *shadowing* or *shadow fading*. In general, such a power deviation due to the shadowing effect is formulated as a zero-mean normally (Gaussian) distributed random variable $X_\sigma$ (in dB) with a standard deviation $\sigma$. With the shadowing effect, the net path loss becomes:

$$PL(d) = \overline{PL}(d) + X_\sigma = 20\log(d) - 20\log(\lambda) - 10\log(A) + X_\sigma. \qquad (2.4)$$

By (2.4), the received power with the same distance $d$ may be different, and has a log-normal distribution. Thus, we also refer to it as the *log-normal shadowing*.

- *Multi-path* is the propagation phenomenon that results in radio waves reaching the receiving antenna by two or more paths. This is mainly caused by reflection and scattering in radio propagation. Since each of these reflected waves takes a different path, it may have a different amplitude and phase. Depending on the phases of these reflected waves, these signals may result in increased or decreased received power at the receiver. Even a very slight change in the propagation path may result in a significant difference in phases of the signals, and therefore in the total received power. Thus, the multi-path propagation will lead to the fast fluctuation of radio wave. Furthermore, multiple versions of radio wave may arrive at the receiving antenna at different times. In this way, the multi-path propagation will introduce a delay spread into the radio wave. This may impose server interference on the successive signals, which is usually called the inter-symbol interference (ISI).

For clarity, we illustrate the above three components of channel response in Figure 2.1. The gray dot dashed curve represents the distance-dependent path loss (without shadowing). The red dashed curve represents the path loss with log-normal shadowing, where the log-normal shadowing makes the total path loss deviates from the distance-dependent one. The blue solid curve represents the real channel response with the path loss, shadowing, and multi-path propagation. Note that the variations due to the multi-path propagation change at distances in the scale of wavelength.

## 2.2    WIRELESS MULTIPLE ACCESS TECHNOLOGIES

Wireless multiple access technology is very important for modern communication systems, as it allows multiple users to share the limited communication resources. In wireless communication systems, multiple access technologies are usually based on *multiplexing*. In this section, we will discuss several widely used multiple access schemes, including the frequency division multiple access (FDMA), orthogonal frequency division multiple access (OFDMA), time division multiple access (TDMA), code division multiple access (CDMA), and random access technologies such as carrier sense multiple access (CSMA).

### 2.2.1    FDMA AND OFDMA TECHNOLOGIES

The frequency division multiple access (FDMA) channel access scheme is based on the frequency division multiplex technology, which provides different frequency bands to different mobile users.

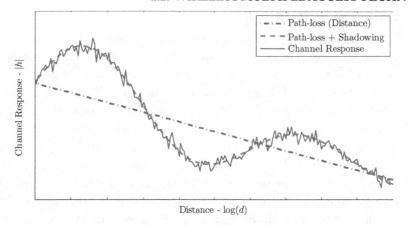

**Figure 2.1:** Illustrations of path loss, shadowing, and multi-path propagation.

That is, it allows several users to transmit at the same time by using different frequency bands. Typical FDMA systems include the second-generation (2G) cellular communication systems such as Global System for Mobile Communications (GSM), where each phone call is assigned to a specific uplink channel and a specific downlink channel.

An advanced form of FDMA is the orthogonal frequency division multiple access (OFDMA), which is used in the fourth-generation (4G) cellular communication systems and wireless local area networks (WLAN) based on the latest versions of 802.11 standards. The key feature of OFDMA is that the frequency bands are partially overlapped (but logically orthogonal), and therefore the spectrum efficiency can be greatly improved comparing with FDMA. In an OFDMA system, each mobile user is allowed to use one or multiple channels (more often called sub-carriers), making it flexible to provide different quality of service (QoS) guarantees to different users.

## 2.2.2 TDMA TECHNOLOGY

The time division multiple access (TDMA) channel access scheme is based on the time division multiplex technology, which provides different time slots to different mobile users in a cyclically repetitive frame structure. That is, the whole time period is divided into multiple time slots, each for a particular mobile user. The users transmit in rapid succession, one after the other, each using its own time slot. TDMA has been used in the second-generation (2G) cellular communication systems such as GSM. More precisely, GSM cellular systems are based on the combination of TDMA and FDMA, where each frequency channel is divided into multiple time slots, each carrying one phone call or signaling data.

### 2.2.3   CDMA TECHNOLOGY

The code division multiple access (CDMA) scheme is based on the *spread spectrum* technology, which allows several mobile users to send information simultaneously over a single frequency channel. The key idea of CDMA is to assign each user a different *spreading code*, based on which the signals of multiple users can be separated. The most common form of CDMA is the direct sequence spread spectrum (DS-CDMA), which has been used in the third-generation (3G) cellular communication systems. In this case, each information bit (of a mobile user) is spread to a long code sequence of several pulses, called chips. Such a code sequence is usually referred to as the spreading code. The separation of the signals of multiple users is made by *correlating* the received signal with the locally generated spreading code of the desired user. If the signal matches the desired user's code, then the correlation function will be high and the system can extract that signal. If the desired user's code has nothing in common with the signal, the correlation should be as close to zero as possible (thus eliminating the signal). This process is usually referred to as cross correlation. Obviously, the spreaded signal (chip) has a much higher data rate (bandwidth) than the original data, and thus CDMA is essentially a form of spread-spectrum technology.

### 2.2.4   RANDOM ACCESS TECHNOLOGY

In the previous channel access schemes, each mobile user accesses the transmission medium under the full control of a controller. For example, in CDMA, each user spreads its data by using the spread code assigned by the controller; in TDMA or FDMA, each user occupies the time slot or frequency band assigned by the controller.

In the random access scheme, however, each user has the right to access the medium without being controlled by any other controller. Obviously, if more than one user tries to send data at the same time, there is an access conflict (called a *collision*), and the signals will be either destroyed or polluted. Therefore, it is essential to avoid the access conflict or to resolve it when it happens. The simplest random access scheme is the so-called ALOHA random access, which allow mobile users to initiate their transmissions at any time. Whenever a collision occurs, a mobile user will wait for a *random* time and then try resending the data. A more advanced random access scheme is the carrier sense multiple access (CSMA), in which a mobile user checks the existence of other users' signals before transmitting on a shared transmission medium. That is, it tries to detect the presence of radio waves from other users before attempting to transmit its own data. In other words, CSMA is based on the principle "sense before transmit" or "listen before talk."

## 2.3   WIRELESS NETWORKS

Depending on the transmission range or coverage area, wireless communication networks can be categorized into the following types: wireless personal area network (e.g., IEEE 802.15 Bluetooth), wireless local area network (e.g., IEEE 802.11 WiFi), wireless metropolitan area network (e.g., IEEE 802.16 WiMAX), wireless wide area network (e.g., IEEE 802.20 MobileFi, and 3GPP Cellular), and wireless regional area network (e.g., IEEE 802.22). Wireless communication networks can also be categorized by the access and networking technologies: wireless cellular network, wireless ad-hoc network, wireless sensor network, wireless mesh network, and cognitive radio network. In this section, we will discuss several widely used wireless networks briefly.

## 2.3.1  WIRELESS CELLULAR NETWORK

In a cellular network, a wide geographic area to be covered by radio services is divided into regular shaped zones called *cells*, which can be hexagonal, square, circular or some other regular shapes. Each cell is associated with a fixed-location transceiver, called the base station, which is usually located in the center of the cell. Mobile cellular users are connected with each other via the base stations.

Each cell serves those mobile cellular users within its coverage area via the corresponding base station. Since mobile cellular users can move between cells, thus handoff and mobility management are very important for a cellular network. Although each cell can only provide radio service in a small area, when many cells are joined together they can provide radio coverage over a wide geographic area. A simple illustration of the communication between two mobile users in a cellular network is as follows: each user transmits/receives radio signal to/from the respective base stations, and two base stations are connected through a wired network called core network or backbone network.

To avoid the interference from signals from other cells, the adjacent neighboring cells are usually operated on different frequency bands. On the other hand, to improve the spectrum efficiency, the same frequency band is usually used by multiple cells as long as these cells are far enough apart such that the radio signal of one cell does not cause harmful interferences on other cells. This is the so-called *frequency reuse*.

Today, wireless cellular networks have been widely used in practice. The most common example is the mobile phone network. Depending on the different access technologies, there are a number of different widely used mobile phone systems, including: global system for mobile communications (GSM), general packet radio service (GPRS), enhanced data rates for GSM evolution (EDGE), and universal mobile telecommunications system (UMTS). Most of these cellular systems are based on the 3GPP standards.

## 2.3.2  WIRELESS LAN NETWORK

A wireless local area network (wireless LAN, or WLAN) is usually used to provide high-speed radio service in a local small area. The most common architecture of a WLAN system is based on an infrastructure-based controller called an *access point*, which is usually connected to a wired network for receiving incoming and sending outgoing traffic. Mobile users communicate with each other or connect to the wider Internet via these access points. In this sense, a WLAN is very similar to a cellular network. Another common architecture of WLAN is the so-called *ad-hoc network*, where each mobile user transmits data to another user directly. Due to the limitation of access technologies, the coverage area of WLANs is usually small (e.g., fewer than 200 meters for each access point). Thus, the mobility of mobile users in a WLAN is rather limited.

Despite of the limited mobility, WLAN has become very popular today due to its ability to provide high-speed communication service. Almost all the smartphones, pads, and laptops are equipped with the WLAN interface. Most modern WLANs are based on the IEEE 802.11 standard, marketed under the Wi-Fi brand name.

## 2.3.3   WIRELESS AD-HOC NETWORK

A wireless ad-hoc network is a type of decentralized wireless network, usually based on the IEEE 802.11 standard. That is, it does not rely on the preexisting infrastructure such as the base station in a cellular network or the access point in a WLAN. Due to the limited transmission range of mobile nodes, a source node may need to communicate to a destination node in a *multi-hop* fashion. Moreover, each node participates in routing by forwarding data for other nodes, and the decision of which nodes forward data is made dynamically based on the network status. Since the nodes are mobile, the network topology may change rapidly and unpredictably over time. Therefore, the nodes need to self-organize to establish network connectivity to support various mobile applications.

The decentralized nature of ad-hoc network makes it suitable for a variety of applications where the centralized control cannot be achieved. It can improve the scalability of networks compared to centralized managed networks such as cellular networks. Moreover, it can be applied to emergency situations like natural disasters or military conflicts due to the minimal need of configuration and quick deployment. The presence of dynamic and adaptive routing protocols enables an ad-hoc network to be formed quickly.

## 2.3.4   WIRELESS SENSOR NETWORK

A wireless sensor network consists of a set of spatially distributed autonomous sensors. These sensors are usually designed to monitor physical or environmental conditions (e.g., temperature, sound, and pressure), and to cooperatively deliver their measured data to a specific location called the *sink node*, through ad-hoc communications. The modern sensor networks are usually bi-directional. That is, they cannot only collect data from sensors passively, but also actively control these sensors. The development of wireless sensor networks was primarily motivated by military applications such as battlefield surveillance. Today, wireless sensor networks have been used in many industrial and consumer applications, such as industrial process monitoring and machine health monitoring.

An important feature of wireless sensor networks is the energy constraint. Specifically, due to the requirements of small size and low cost, the sensors are usually hardware-constrained, and with limited capacity of energy storage (e.g., battery). Moreover, due to the spatial distribution of sensor nodes, it is hard to charge these sensor nodes online and in real-time. For these reasons, the energy resource, in addition to the radio resource, becomes an very important factor in designing wireless sensor network protocols.

## 2.3.5    WIRELESS MESH NETWORK

A wireless mesh network is a communications network made up of radio nodes organized in a mesh topology. It often consists of two kinds of different nodes: mesh clients and mesh routers. The mesh clients are often laptops, cell phones, and other wireless devices, which transmit/receive data to/from other clients or the wider Internet. The mesh routers are often stationary nodes such as base stations or access points, which forward a mesh client's traffic to/from other clients or the gateways which connect to the Internet. Wireless mesh networks can be built upon various wireless technologies such as IEEE 802.11, 802.15, 802.16, and cellular technologies. In an IEEE 802.11-based wireless mesh network, for example, the access points act as mesh routers, and form a mesh backbone for relaying the traffic of mobile users (i.e., mesh clients).

In general, a wireless mesh network can be seen as a special case of the wireless ad-hoc network. In this sense, the mesh routers may be mobile nodes themselves in an ad-hoc network, and can be moved according to specific demands arising in the network. Often the mesh routers are not limited in terms of resources compared to other nodes in the network, and thus can be exploited to perform more resource intensive functions.

## 2.3.6    COGNITIVE RADIO NETWORK

Cognitive radio network is a novel network architecture based on advanced wireless technologies such as cognitive radio and dynamic spectrum access.

Cognitive radio is an adaptive, intelligent radio technology that can automatically detect available frequency bands in a certain frequency range (usually licensed to some organizations or commercial companies); it also enables devices to access the frequency bands distributed in a wide frequency range. Cognitive radio uses a number of technologies, including the adaptive radio where the communications system monitors and modifies its own performance, and the software defined radio (SDR) where traditional hardware components including mixers, modulators, and amplifiers have been replaced with intelligent software.

Dynamic spectrum access is a new paradigm for utilizing wireless spectrum, and has its theoretical roots in network information theory, game theory, machine learning, artificial intelligence, etc. The key idea of dynamic spectrum access is to allow unlicensed devices to access the frequency bands (licensed to other licensees) in an opportunistic manner, whenever such a secondary access does not generate harmful interference to the licensees. Therefore, the reliable detection of the presence of licensed devices is very critical for dynamic spectrum access. Early studies on licensed devices detection mainly focused on the spectrum sensing techniques. However, recent studies showed that the pure spectrum sensing is often inadequate or inefficient, due to the high deployment cost and low detection performance. As a consequence, many regulatory bodies (e.g., FCC in the US and Ofcom in the UK) and standardization organizations (e.g., IEEE and ECC) have been advocating an alternative solution, which relies on a centralized third-party database called *geo-location database*. Specifically, the geo-location database maintains the up-to-date spectrum usage information of licensees, and can identify the available frequency bands at a particular time and place for secondary access.

## 2.4   RADIO RESOURCE MANAGEMENT

A fundamental issue of the operation of wireless networks is the radio resource management (RRM), which provides the system level control of interference, efficiency, and other transmission characteristics in wireless communication systems. It usually involves strategies and algorithms for controlling network parameters such as transmit power, channel allocation, data rates, handover criteria, modulation scheme, error coding scheme, etc. The objective of RRM is to utilize the limited radio spectrum resources and radio network infrastructures as efficiently as possible. In this section, we will discuss the most common RRM problems in wireless communication systems.

### 2.4.1   POWER CONTROL

Transmit power control is one of the most important issues in wireless communication systems. Broadly speaking, power control is the intelligent selection of transmit power so as to achieve a good system performance (e.g., low mutual interference, high network capacity, and wide geographic coverage area). Power control has been widely used in many types of wireless networks, including wireless cellular networks, sensor networks, wireless LANs, etc.

Power control is particularly important for a CDMA system, where multiple mobile users send information simultaneously over a single frequency channel using different spread codes. Because of this, the transmission of one user will inevitably cause interference on other users' transmissions. Power control in a CDMA system can help to achieve an efficient utilization of the energy resource, and effectively reduce the mutual interferences between mobile users.

Power control is also important for a cellular network based on the FDMA technology (e.g., GSM, GPRS, and EDGE). Although mobile users within the same cell do not interfere with each other (since they use different frequency bands), they may generate interference to users in other cells due to frequency reuse. That is, multiple (non-adjacent) cells may use the same set of frequency bands, and thus the mobile users in these cells may be assigned to the same frequency band. Although the interference of these mobile users can be largely mitigated by the distance factor, there may still be certain remaining interference. Therefore, the joint control of users' (or cells') transmit powers is important to further reduce the interference and improve the network performance.

### 2.4.2   CHANNEL ALLOCATION

In wireless communication systems and especially in a wireless cellular network, channel allocation schemes are required to allocate frequency bands (or channels) to base stations, access points, and mobile devices. The objective is to achieve a high spectrum efficiency by means of frequency reuse, under the constraints of co-channel interference and adjacent channel interference among nearby cells or networks that share the spectrum band.

There are usually two types of different channel allocation strategies: fixed channel allocation (FCA) and dynamic channel allocation (DCA). In the former case, each cell is given a pre-determined set of channels. That is, the number of channels in every cell remains constant irrespective of the number of users in that cell. This may result in traffic congestion in some cells, while a waste of resource in other cells. In the latter case, channels are not allocated to cells permanently; but instead, cells request channels dynamically based on their real-time traffic load. DCA allows the number of channels in a cell to vary with the traffic load, and thus it can usually achieve a higher network capacity.

Another type of important channel allocation problem is the *sub-channel allocation* problem in an OFDMA system (e.g., 4G cellular network). In an OFDMA system, each mobile user is assigned to a sub-carrier or a sub-channel (i.e., a group of sub-carriers) for their transmissions. Due to the heterogeneity of mobile users, the same sub-channel may have different wireless characteristics (e.g., channel responses) for different users. Therefore, the assignment of all sub-channels among all mobile users becomes a challenging problem. Ideally, every sub-channel will be allocated to a mobile user who has a good channel response on this sub-channel. This is the so-called *multiuser diversity*.

### 2.4.3 ADMISSION CONTROL

Admission control is important for wireless communication systems, especially those with limited resources but many potential users. However, an overly conservative admission control policy may reject too many users and result in the under-utilization of radio resources.

Admission control can also be used to differentiate mobile users according to their QoS requirements. For example, voice call traffic usually has a strict QoS requirement (e.g., delay and bandwidth), while data traffic usually has a more flexible QoS requirement. Thus, voice traffic may be admitted with a higher priority than data traffic when the network is congested.

## 2.5    CHAPTER SUMMARY

In this chapter, we introduce the basics of wireless communications and networks, including the radio propagation and radio channel model, wireless access and networking technologies, and several issues of radio resource management. This chapter serves as the basis for understanding the technology aspects of wireless pricing models in later chapters. For more in-depth discussions regarding the wireless technologies, we refer the readers to [8, 9, 10, 11, 12, 13, 14, 15].

CHAPTER 3

# Economics Basics

In this chapter, we will follow the convention of economics and use the terms "firm" and "consumer." A firm may represent a wireless service provider or a wireless spectrum owner, and a consumer can represent a wireless user or a lower tier wireless service provider. In later chapters, we will give more concrete examples of firms and consumers in different wireless networks. The theory introduced in this chapter closely follows several microeconomics textbooks including [16, 17, 18].

## 3.1 SUPPLY AND DEMAND

Supply and demand in a market are both functions of market prices. When prices increase, usually the market supply increases as firms have more incentives to produce, and market demand decreases as consumers have fewer incentives to purchase. We first study how the demand and supply change with prices, and then characterize what prices lead to a market equilibrium where supply equals demand.

### 3.1.1 MARKET DEMAND FUNCTION

Let us consider a consumer who subscribes to a wireless cellular data plan. We may characterize the consumer demand as a function of price by the following table:

**Table 3.1:** Relationship between the monthly wireless data demand and the price per Gigabyte

| Price Per Gigabyte | Wireless Data Demanded Per Month |
|---|---|
| $1 | 50 Gigabytes |
| $2 | 22 Gigabytes |
| $10 | 4 Gigabytes |
| $20 | 1.5 Gigabytes |

Other consumers may have different demands for wireless data. If we add up all consumers' demands together, we will obtain the relationship between the aggregate demand and the price, which we call the market demand function.

**Definition 3.1** The **market demand function** $D(\cdot)$ characterizes the relationship between the total demand quantity $Q_d$ and the product price $P$ as follows

$$Q_d = D(P). \tag{3.1}$$

Figure 3.1 gives an example of the market demand function. Here we adopt the convention of placing price at the vertical $y$-axis and quantity (e.g., demand or supply) at the horizontal $x$-axis. When the price decreases from $P_1$ to $P_2$, the demand increases from $Q_1$ to $Q_2$. There are two reasons for this inverse change of demand. First, the existing consumers who have positive demands at price $P_1$ will increase their demands when the price drops. Second, some consumers did not purchase at price $P_1$ may decide to purchase at the lower price $P_2$.

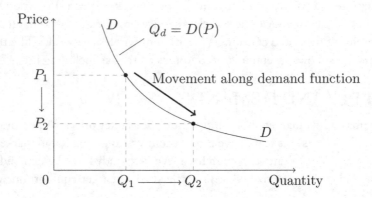

**Figure 3.1:** The market demand function $Q_d = D(P)$. When the price decreases from $P_1$ to $P_2$, the demand increases from $Q_1$ to $Q_2$.

Besides shifting *along* the demand function due to the price change, the demand function itself might also shift due to several reasons: (i) the change of consumers' income, (ii) the price change of other products, and (iii) the change of consumers' tastes. Figure 3.2 illustrates such an example. Let us take wireless data service as an example. When consumers' income increases, the aggregate cellular data demand will increase (and thus the demand function will shift to the right), as consumers are more willing to use high price services (such as high-definition video streaming). When the price of a substitutable product (such as the price of commercial Wi-Fi access points) decreases, the aggregate cellular data demand decreases, as consumers are more willing to use the substitutable product. Finally, when consumers' tastes change due to education, the demand function may also shift.

### 3.1.2  MARKET SUPPLY FUNCTION

Much like Definition 3.1, we can define the market supply function as follows.

**Definition 3.2**    The **market supply function** $S(\cdot)$ characterizes the relationship between the total supply quantity $Q_s$ and the product price $P$ as follows

$$Q_s = S(P). \tag{3.2}$$

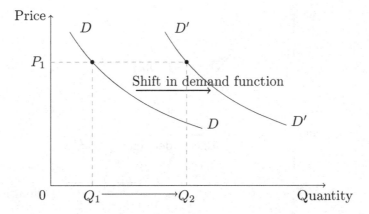

**Figure 3.2:** The shift of market demand function from $Q_d = D(P)$ to $Q'_d = D'(P)$. For example, under the same price $P_1$, the demand changes from $Q_1$ to $Q_2$.

Imagine the case where each firm does not have a capacity limit, and then the total market supply will increase with the price, as shown in Figure 3.3.

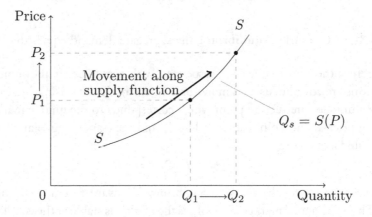

**Figure 3.3:** The market supply function $Q_s = S(P)$ and the shift along the function due to a price increase. For example, when the price increases from $P_1$ to $P_2$, the supply increases from $Q_1$ to $Q_2$.

Similarly, the market supply function itself may shift when the price of a raw material (used for production) changes or the production technology changes. For example, consider a wireless service provider selling wireless services (e.g., data rates) to customers. The supply of wireless resource may change if the price for wireless spectrum (raw material that provides data rates) changes or the physical layer technologies change (such as upgrading from the 3G CDMA-based cellular network

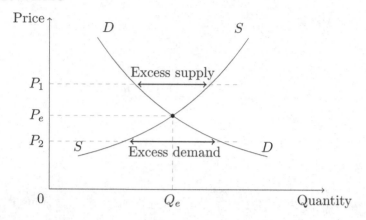

**Figure 3.4:** The market equilibrium price $P_e$ and quantity (demand and supply) $Q_e$.

to a more efficient 4G OFDMA-based network). We leave it as an exercise for the readers to draw a figure of shifting market supply function similar to Figure 3.2.

### 3.1.3  MARKET EQUILIBRIUM

Now let us look at the interactions between supply and demand, which lead to a stable market.

**Definition 3.3**    At a **market equilibrium**, the aggregate demand equals the aggregate supply.

Apparently there will be a price associated with a market equilibrium. If the demand and supply functions are continuous, and monotonic (i.e., strictly decreasing/increasing) with the price, then there is a unique intersection point which corresponds to the unique market equilibrium. The corresponding price at this point is denoted by $P_e$ and the (same) aggregate demand and aggregate supply is denoted as $Q_e$, i.e.,

$$Q_e = D(P_e) = S(P_e). \tag{3.3}$$

Figure 3.4 illustrates the market equilibrium. We want to emphasize that equilibrium is a prediction of how the actual market will look, as the market is stable at the equilibrium and is unlikely to change once it has reached there. When the market price is lower than the equilibrium price $P_e$, for example, the aggregate demand is higher than the aggregate supply. In this case, consumers are willing to pay more to secure the limited supply, and the firms have incentives to produce more to earn more profits. As a result, the market price increases until the equilibrium is reached. Of course, the real process of reaching the market equilibrium is more complicated than this, and may involve some iterative adjustments of the price.

When either market demand function or market supply function shifts due to factors other than the price, the equilibrium will also change accordingly.

## 3.2   CONSUMER BEHAVIOR

Now let us zoom into the behavior of a particular consumer, and understand how the market demand function $Q_d = D(P)$ is derived.

### 3.2.1   INDIFFERENCE CURVES

In order to understand a single consumer's demand, we first need to understand how a consumer evaluates the benefit of consuming certain products. For example, how would a consumer evaluate the satisfaction level of watching a 60-minute action movie and playing 30 minutes of video games on his iPad? To explain this, we first define the concept of market basket (also known as the commodity bundle).

**Definition 3.4**   A **market basket** specifies the quantity of different products.

If we consider "watching movies" and "playing games" as two types of products, then watching a 60-minute movie and playing 30 minutes of a game can be represented by the market basket (60, 30). We can use a utility function $U$ to characterize the consumer's satisfaction level of consuming a certain market basket $(x, y)$, i.e.,

$$U = U(x, y). \tag{3.4}$$

In Figure 3.5, we represent the basket (60, 30) as point 1. We also add several baskets, where basket 2 is (45, 40), basket 3 is (30, 60), basket 4 is (25, 25), and basket 5 is (75, 65). Assuming that the consumer's utility is increasing in both $x$ and $y$, then point 5 leads to the maximum utility (among five baskets) and point 4 leads to the minimum utility. If we further know that the consumer is indifferent among baskets 1, 2, and 3 (i.e., the consumer associates these baskets with the same utility value), then we say that these three baskets are on the same indifference curve.

**Definition 3.5**   An **indifference curve** represents a set of market baskets where the consumer's utilities are the same.

The indifference curve characterizes how a consumer trades off two different products. We can further imagine an indifference map, which consists of all indifference curves of a consumer. In Figure 3.5, basket 5 will be on an indifference curve that has a higher utility than baskets 1, 2, and 3, and basket 4 will be on an indifference curve that has a lower utility than baskets 1, 2, and 3.

### 3.2.2   BUDGET CONSTRAINTS

If a consumer has enough income, he will definitely prefer to choose basket 5 in Figure 3.5 over the other four baskets. However, the budget constraint will limit a consumer's choice.

**Definition 3.6**   The **budget constraint** characterizes which market baskets are affordable to the consumer.

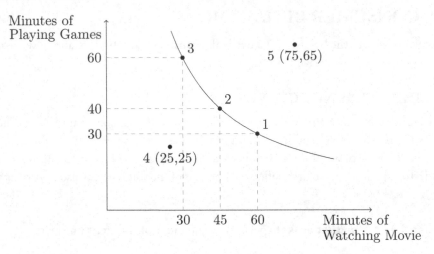

**Figure 3.5:** Market baskets and indifference curve.

In our example, we can consider the limited energy of the iPad battery as the budget. Assuming watching one minute of movie will cost 1 unit of energy, and playing one minute of game will cost 2 units of energy. Then the constraint of 100 units of energy leads to the budget constraint shown in Figure 3.6, which can be mathematically represented as $x + 2y = 100$, where $x$ and $y$ are the times for watching movie and for playing game, respectively. The consumer can afford any market basket on or below the budget constraint. Alternatively, one can think of the price of watching movie as $P_x = 1/\text{min}$ and the price of playing game as $P_y = 2/\text{min}$. Thus, the budget constraint can be represented by $x P_x + y P_y = I$, where $I$ is the fixed budget.

## 3.2.3   CONSUMER CONSUMPTION PROBLEM

Once we consider both the consumer's indifference curve and the budget constraint, we will start to understand how a consumer decides which market basket to purchase. Essentially, the consumer wants to maximize its utility subject to the budget constraint. Geometrically, the consumer will find the highest indifference curve that "touches" the budget constraint.

Let us consider the illustration in Figure 3.7. It is clear that basket $a$ or $b$ does not maximize the utility, as basket $c$ is on a higher utility indifference curve which "touches" the budget constraint (and thus is feasible). To be more precise, the derivative of the indifference curve with utility $U_3$ at basket $c$ equals the slope of the budget constraint at basket $c$, i.e., the budget constraint is the tangent line to the indifference curve at basket $c$,

$$\left. \frac{\Delta y}{\Delta x} \right|_{U(x,y)=U_3,(x,y)=(x_c,y_c)} = -\frac{P_x}{P_y}. \tag{3.5}$$

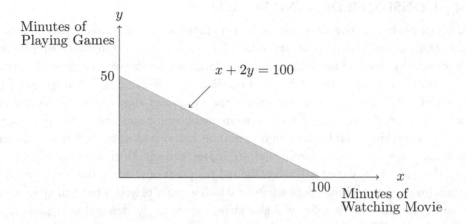

**Figure 3.6:** Budget constraint of 100 units of battery energy: $x + 2y = 100$. The consumer can afford any market basket on or below the budget constraint (i.e., the shaded area).

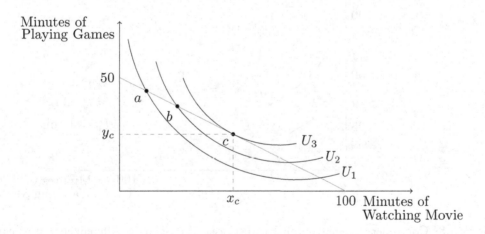

**Figure 3.7:** Consumer's optimal market basket choice is basket $c$.

The left-hand side of equation (3.5) is also called **marginal rate of substitution (MRS)**, which represents how much the consumer is willing to tradeoff one product with the other product. Here we constrain $U(x, y) = U_3$, which means that the MRS is measured along the indifference curve with a constant utility $U_3$. Except in very special cases, the MRS along an indifference curve is not a constant, and that is why we need to specify basket c at $(x, y) = (x_c, y_c)$.

### 3.2.4 CONSUMER DEMAND FUNCTION

Now we are ready to derive a consumer's demand function, which characterizes how its demand of a product changes with the price of that product. The market demand function is simply the summation of all consumers' demand functions in the same market (also known as aggregate demand).

Assume that there are three games on iPad. The first one is a strategy game (e.g., Chess) that requires deep thinking and thus infrequent inputs and animations; the second one is a light game (e.g., Angry Birds) that contains some frequent simple animations; the third one is an action game (e.g., car racing) that features high-definition action-packed animations. The energy prices of these three games are 1/min, 2/min, and 4/min, respectively. With a total budget of 100 units of energy, the budget constraint will rotate around the point of (100, 0) (under the fixed energy price of 1/min for movie watching), depending on which game is played. The optimal market basket that maximizes the consumer's utility will also change accordingly, denoted as baskets A, B, and C as shown in Figure 3.8.

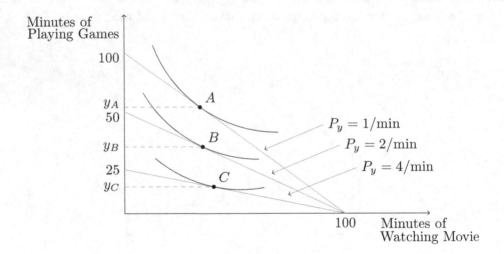

**Figure 3.8:** Consumer's different optimal market basket choices under different energy prices for playing games.

The three points of A, B, C lead to three points on the consumer's demand curve (in terms of the demand of time for playing game). Connecting these points (or alternatively choosing different energy price for playing games and examining the utility maximizing baskets) will lead to the demand function as shown in Figure 3.9.

### 3.2.5 PRICE ELASTICITY

We notice that a consumer's demand is often downward slopping, i.e., a lower price leads to a higher demand. However, how fast the demand changes with the price depends on the nature of

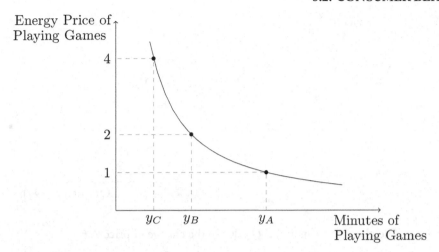

**Figure 3.9:** Consumer's demand function for playing iPad games as a function of the energy cost.

the demand. Consider the cellular wireless data usage as an example. A college student might be very price sensitive, and will dramatically decrease the monthly data usage if the price per Gigabyte of cellular data increases. However, a business consumer might be much less sensitive and not even notice the change of price until several months later. Such sensitivity of demand in term of price can be characterized by the price elasticity.

**Definition 3.7**   The **price elasticity of demand** measures the ratio between the percentage change of demand and the percentage change of price, i.e.,

$$E_d = \frac{\% \text{ change in demand}}{\% \text{ change in price}} = \frac{\Delta Q_d / Q_d}{\Delta P / P}. \tag{3.6}$$

An illustrative example is shown in Figure 3.10. Here we use the market demand function $Q_d = D(P)$ to illustrate the concept of price elasticity, although the same concept can also be applied to consumer demand function. The value of $E_d$ is often negative due to the downward slopping of the demand curve.

When the demand function $Q_d$ is differentiable, we can compute the "point-price elasticity" by taking derivative of the demand function at a particular price $P$:

$$E_d = \frac{P}{Q_d} \frac{\partial Q_d}{\partial P}. \tag{3.7}$$

Depending on the value of $E_d$, the demand can be classified into three types:

- *Elastic demand:* the demand changes significantly with the price and $E_d < -1$.

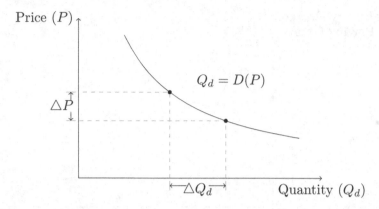

**Figure 3.10:** The change of demand $\Delta Q_d$ due to the change of price $\Delta P$.

- *Inelastic demand:* the demand is not sensitive to price and $-1 < E_d < 0$.

- *Unitary elastic demand:* $E_d = -1$.

Notice that different parts of the same demand function can have different price elasticities. If a firm can adjust the price $P$ to maximize its revenue $P Q_d$, then it will decrease the price when the market demand is elastic, increase the price when the market demand is inelastic, and do not change the price when the market demand is unitary elastic.

## 3.3   FIRM BEHAVIOR

In this section, we will take a deeper look at the firm, and discuss how the market supply function $Q_s = S(P)$ is derived from the firm's cost minimization behavior.

### 3.3.1   TOTAL AND MARGINAL PRODUCTION COST

In a market, a firm will produce products based on certain technologies and sell the products in the market. How many they produce depends both on the production costs and the selling price in the market. We will start by understanding the types and impacts of production costs.

First, we can classify the cost into *explicit costs* and *opportunity costs*. Explicit cost of a wireless service provider may involve the cost of purchasing and installing the network equipments as well as the salary of the engineers. Opportunity costs represent the income that the firm loses due to utilizing the resources for a particular purpose. For example, if a spectrum owner (such as AT&T) decides to offer cellular services over its licensed spectrum, it explicitly forgoes the income that it can earn by leasing the spectrum to a third party (such as Google). The production cost thus includes both the explicit and opportunity costs.

The production cost will be different depending on whether we consider *short-term* or *long-term*. In general, we have fewer production choices in the short run than in the long run. For example, in the long run a wireless service provider may be able to choose which technology to use (CDMA, TDMA, or OFDMA) and how much spectrum to obtain (through auction or leasing). In the short run, however, both the technology and total spectrum (and thus the network capacity) are fixed, and the service provider can only change the resource allocation among different cells, users, frequency bands, and time slots. In this section, we will focus the discussions on the short-term production cost. The discussions can be similarly generalized to the long-term production cost.

The total cost includes two parts: the *fixed* cost and the *variable* cost. The fixed cost $F$ is the amount that a firm needs to pay independent of the quantity produced. The variable cost $V(q)$ depends on the production quantity $q$.

**Definition 3.8**  The **total production cost** includes both the fixed cost and variable cost, i.e.,

$$C(q) = F + V(q). \tag{3.8}$$

We are also interested in how the total cost changes when the firm changes with production quantity.

**Definition 3.9**  The **marginal cost** measures how the total cost changes with the production quantity, i.e.,

$$MC(q) = \frac{\%\text{ change in total production cost}}{\%\text{ change in production quantity}} = \frac{\Delta C(q)}{\Delta q} = \frac{\Delta V(q)}{\Delta q}. \tag{3.9}$$

Notice that the fixed cost $F$ does not affect the computation of marginal cost. When the variable cost function $V(q)$ is differentiable, we have

$$MC(q) = \frac{\partial C(q)}{\partial q} = \frac{\partial V(q)}{\partial q}. \tag{3.10}$$

## 3.3.2  COMPETITIVE FIRM'S SUPPLY FUNCTION

Next we derive the supply function of a *competitive* firm.

**Definition 3.10**  A **competitive firm** is price-taking and acts as if the market price is independent of the quantity produced and sold by the firm.

The competitive firm accurately reflects the reality when the firm faces many competitors in the same market. In this case, each firm's production decision is unlikely to significantly change the total quantity available in the market, and thus will not significantly affect the market price. The total revenue of a competitive firm will be $P \cdot q$, where $P$ is the market price and $q$ is the production quantity. This is assuming that the produced quantity can always be sold at the fixed market price $P$. The firm wants to choose the production amount $q$ to maximize its profit.

**Definition 3.11**    A competitive firm's **profit** is the difference between revenue and total cost, i.e.,

$$\pi(q) = P \cdot q - V(q) - F. \tag{3.11}$$

If the firm produces $q = 0$, then the total profit is $-F$. Here we assume that the fixed cost $F$ is also the *sunk cost*, i.e., a cost that the firm cannot avoid. This means that a firm will only produce when the revenue is no less than the variable cost, i.e., $Pq \geq V(q)$. At the optimal choice of $q^*$ that maximizes the profit, we have

$$P = \frac{\partial V(q)}{\partial q} = MC(q), \tag{3.12}$$

which means that the price equals the marginal cost.

As we change the market price $P$, the competitive firm's optimal production quantity $q$ changes according to (3.12). The firm's supply function is thus the firm's marginal cost function as long as revenue is no smaller than the variable cost.

## 3.4    CHAPTER SUMMARY

In this chapter, we introduced the basics of microeconoomics, including the relationship of supply and demand, the consumer behavior model, and the firm behavior model. In particular, we showed how the market supply and demand are derived based on the behaviors of individual consumers and competitive firms. This chapter serves as the basis for understanding the economics aspects of wireless pricing models in later chapters.

# CHAPTER 4

# Social Optimal Pricing

This chapter will focus on the issue of social optimal pricing, where a service provider chooses prices to maximize the social welfare. This corresponds to the case, for example, where the service provider's interests are aligned with the regulator's interests through proper economic mechanisms. The basic approach of social optimal pricing is to formulate the problem as an optimization problem, and design a dual-based distributed algorithm for the distributed resource allocation. Here the dual variables have the interpretation of "shadow prices" in economics.

We will first introduce the theoretical background of convex optimization and dual-based algorithms, and then illustrate the theory through two examples: single cell wireless video streaming and multi-provider resource allocation.

## 4.1 THEORY: DUAL-BASED OPTIMIZATION

In this section, we will cover the basics of convex optimization and dual-optimization. We closely follow the discussions in [19, 20], where readers can find more in-depth discussions.

### 4.1.1 PRELIMS

We use the notation $\mathbb{R}^n$ to denote the set of all real $n$-vectors. Each vector in $\mathbb{R}^n$ is called a *point* of $\mathbb{R}^n$. When $n = 1$, we use $\mathbb{R}$ to represent the set of real 1-vectors or real numbers. The notation $f : \mathbb{R}^n \to \mathbb{R}^m$ is used to denote a function on some *subset* of $\mathbb{R}^n$ (specifically, its *domain*, which we denote $\mathcal{D}(f)$) into the set $\mathbb{R}^m$. That is, a function $f : \mathbb{R}^n \to \mathbb{R}^m$ maps every real $n$-vector *in its domain* $\mathcal{D}(f)$ into an $m$-vector.

**Convex Sets**

Suppose $x_1 \neq x_2$ are two distinct points in $\mathbb{R}^n$. Any point $y$ on the *line* passing through $x_1$ and $x_2$ can be expressed as

$$y = \theta x_1 + (1 - \theta)x_2, \quad \text{for some } \theta \in \mathbb{R}.$$

The parameter value $\theta = 1$ corresponds to $y = x_1$, and $\theta = 0$ corresponds to $y = x_2$. Values of $\theta$ between 0 and 1 correspond to the (closed) *line segment* between $x_1$ and $x_2$.

A point on the line passing through $x_1$ and $x_2$ is referred to as an *affine combination* of $x_1$ and $x_2$. A point on the line segment between $x_1$ and $x_2$ is referred to as a *convex combination* of $x_1$ and $x_2$, which can be equivalently expressed as

$$y = \theta_1 x_1 + \theta_2 x_2,$$

with $\theta_1 + \theta_2 = 1$ and $\theta_i \geq 0, i = 1, 2$.

A nonempty set $\mathcal{X} \subseteq \mathbb{R}^n$ is *convex* if the line segment between any two points (i.e., convex combinations of any two points) in $\mathcal{X}$ lies entirely in $\mathcal{X}$. Specifically,

**Definition 4.1  Convex Set**  A nonempty set $\mathcal{X} \subseteq \mathbb{R}^n$ is *convex* if for any $\boldsymbol{x}_1, \boldsymbol{x}_2 \in \mathcal{X}$ and any $\theta \in \mathbb{R}$ with $0 \leq \theta \leq 1$, we have

$$\theta \boldsymbol{x}_1 + (1 - \theta)\boldsymbol{x}_2 \in \mathcal{X}. \tag{4.1}$$

Geometrically, a set is convex if every point in the set can be reached by every other point, along an *inner straight path* between them, where inner means lying in the set. Obviously, any interval in $\mathbb{R}$ is a convex set. Figure 4.1 illustrates some simple convex and nonconvex sets in $\mathbb{R}^2$.

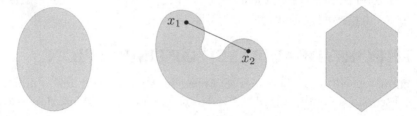

**Figure 4.1:** Some simple convex and nonconvex sets. (I) The ellipsoid, which includes its boundary (shown as solid curves), is convex. (II) The kidney shaped set is not convex, since the line segment between the points $\boldsymbol{x}_1$ and $\boldsymbol{x}_2$ is not entirely contained in the set. (III) The hexagon which contains some boundary points but not all (the dotted boundary points are not included), is not convex.

The concept of convex combination can be generalized to more than two points. Specifically, a convex combination of points $\boldsymbol{x}_1, ..., \boldsymbol{x}_k$ can be expressed as

$$\boldsymbol{y} = \theta_1 \boldsymbol{x}_1 + ... + \theta_k \boldsymbol{x}_k, \tag{4.2}$$

with $\theta_1 + ... + \theta_k = 1$ and $\theta_i \geq 0, i = 1, ..., k$. The condition for convex sets can be generalized accordingly. More specifically,

**Lemma 4.2**  *A nonempty set $\mathcal{X}$ is convex, if and only if for any $\boldsymbol{x}_1, ..., \boldsymbol{x}_k \in \mathcal{X}$,*

$$\theta_1 \boldsymbol{x}_1 + ... + \theta_k \boldsymbol{x}_k \in \mathcal{X}, \tag{4.3}$$

*when $\theta_1 + ... + \theta_k = 1$ and $\theta_i \geq 0, i = 1, ..., k$.*

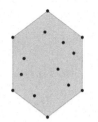 

**Figure 4.2:** The convex hulls of two simple sets in $\mathbb{R}^2$. (I) The convex hull of a set of discrete points (shown as dots) is the pentagon (shown shaded). (II) The convex hull of the kidney shaped set in Figure 4.1 is the shaded set.

The *convex hull* of a set $\mathcal{X}$, denoted $\mathcal{H}(\mathcal{X})$, is the *smallest* convex set that contains $\mathcal{X}$. That is, it consists of the convex combinations of all points in $\mathcal{X}$. Specifically,

**Definition 4.3   Convex Hull**   The *convex hull* of a set $\mathcal{X}$, denoted $\mathcal{H}(\mathcal{X})$, is given by

$$\mathcal{H}(\mathcal{X}) \triangleq \{\theta_1 \boldsymbol{x}_1 + ... + \theta_k \boldsymbol{x}_k \mid \theta_1 + ... + \theta_k = 1, \theta_i \geq 0, \boldsymbol{x}_i \in \mathcal{X}, i = 1, ..., k\}.$$

As the name suggests, the convex hull $\mathcal{H}(\mathcal{X})$ is always convex. Moreover, we have (i) $\mathcal{X} \subseteq \mathcal{H}(\mathcal{X})$, (ii) $\mathcal{X} = \mathcal{H}(\mathcal{X})$ if $\mathcal{X}$ is a convex set, and (iii) If $\mathcal{Y}$ is any convex set that contains $\mathcal{X}$, then $\mathcal{H}(\mathcal{X}) \subseteq \mathcal{Y}$. The last statement implies that the convex hull of a set $\mathcal{X}$ is the smallest convex set that contains $\mathcal{X}$. Figure 4.2 illustrates the convex hulls of some simple sets in $\mathbb{R}^2$.

### Operations Preserving Convexity of Sets

Now we describe some simple operations that preserve the convexity of sets, or allow us to construct new convex sets.

1. *Intersection*: If $\mathcal{X}_1, ..., \mathcal{X}_k$ are convex sets, then $\mathcal{X} \triangleq \mathcal{X}_1 \cap ... \cap \mathcal{X}_k$ is convex.

2. *Affine mapping*: Suppose $\mathcal{X}$ is a subset of $\mathbb{R}^n$, $\mathbf{A} \in \mathbb{R}^{m \times n}$, and $\boldsymbol{b} \in \mathbb{R}^m$. Define a new set $\mathcal{Y} \subseteq \mathbb{R}^m$ by[1]

$$\mathcal{Y} \triangleq \{\mathbf{A}\boldsymbol{x} + \boldsymbol{b} \mid \boldsymbol{x} \in \mathcal{X}\}.$$

   Then if $\mathcal{X}$ is convex, so is $\mathcal{Y}$. The affine mapping operation generalizes a lot of common operations including scaling, translation, summation, projection, etc.

---

[1] Here the notation $\mathbb{R}^{m \times n}$ denotes the set of all $m \times n$ real matrices, and $\mathbf{A}$ is an arbitrary $m \times n$ matrix.

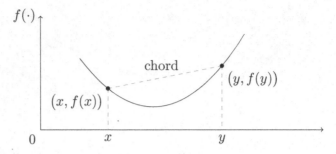

**Figure 4.3:** An illustration of a convex function $f(\cdot)$ on $\mathbb{R}$. The chord (shown as dots) between points $(\boldsymbol{x}, f(\boldsymbol{x}))$ and $(\boldsymbol{y}, f(\boldsymbol{y}))$ on the graph of $f(\cdot)$ lies above the graph.

**Convex Functions**

We consider a scalar-valued function $f : \mathbb{R}^n \to \mathbb{R}$, which maps every real $n$-vectors in its domain $\mathcal{D}(f)$ into a real number in $\mathbb{R}$. In order to distinguish a function from a variable, we will use the notation $f(\cdot)$ to denote a function $f$ whenever there is a need.

In the context of optimization, one of the most important properties of a function $f(\cdot)$ is its convexity (or concavity). Specifically,

**Definition 4.4  Convex Function**   A function $f : \mathbb{R}^n \to \mathbb{R}$ is *convex*, if (i) $\mathcal{D}(f)$ is a convex set, and (ii) for all $\boldsymbol{x}, \boldsymbol{y} \in \mathcal{D}(f)$ and $\theta \in \mathbb{R}$ with $0 \leq \theta \leq 1$, we have

$$f(\theta\boldsymbol{x} + (1 - \theta)\boldsymbol{y}) \leq \theta f(\boldsymbol{x}) + (1 - \theta)f(\boldsymbol{y}). \tag{4.4}$$

A function $f(\cdot)$ is *strictly convex* if the strict inequality holds in (4.4) whenever $\boldsymbol{x} \neq \boldsymbol{y}$ and $0 < \theta < 1$. We say $f(\cdot)$ is (strictly) *concave* if $-f(\cdot)$ is (strictly) convex. Note that a function can be neither convex nor concave. As a simple example, consider the function $f(x) = x^3$ on $\mathbb{R}$. We can easily find that $f(\theta x + (1 - \theta)y) \geq \theta f(x) + (1 - \theta)f(y)$ when $x, y \leq 0$, and $f(\theta x + (1 - \theta)y) \leq \theta f(x) + (1 - \theta)f(y)$ when $x, y \geq 0$.

Geometrically, the inequality in (4.4) means that for any $\boldsymbol{x}, \boldsymbol{y} \in \mathcal{D}(f)$, the line segment between points $(\boldsymbol{x}, f(\boldsymbol{x}))$ and $(\boldsymbol{y}, f(\boldsymbol{y}))$, which is called the *chord* from $\boldsymbol{x}$ to $\boldsymbol{y}$, lies above the graph of $f(\cdot)$. Figure 4.3 illustrates a simple convex function on $\mathbb{R}$.

As with convex sets, the condition in (4.4) can be generalized to the case of more than two points: A function $f(\cdot)$ is convex, if and only if $\mathcal{D}(f)$ is convex and

$$f(\theta_1\boldsymbol{x}_1 + \ldots + \theta_k\boldsymbol{x}_k) \leq \theta_1 f(\boldsymbol{x}_1) + \ldots + \theta_k f(\boldsymbol{x}_k), \tag{4.5}$$

for any $\boldsymbol{x}_1, \ldots, \boldsymbol{x}_k \in \mathcal{D}(f)$, when $\theta_1 + \ldots + \theta_k = 1$ and $\theta_i \geq 0, i = 1, \ldots, k$.

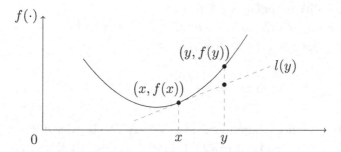

**Figure 4.4:** The first-order condition for a convex function $f(\cdot)$ on $\mathbb{R}$. The line $l(y) = f(x) + \nabla f(x)^T (y - x)$ (shown as a dashed line) lies under the graph of $f(\cdot)$.

**First-order Conditions for Convex Functions**

For a scalar-valued function $f : \mathbb{R}^n \to \mathbb{R}$, the derivative or *gradient* of $f(\cdot)$ at a point $x \in \mathcal{D}(f)$, denoted by $\nabla f(x)$, is an $n$-vector with the $i$th component given by

$$\nabla f(x)_i = \frac{\partial f(x)}{\partial x_i}, \ i = 1, ..., n, \tag{4.6}$$

provided the partial derivatives exist. Here $x_i$ is the $i$-th coordinate of the vector $x$. If the partial derivatives exist at $x$ for all coordinates $x_i$, we say $f(\cdot)$ is *differentiable* at $x$. The function $f(\cdot)$ is differentiable (everywhere in its domain) if $\mathcal{D}(f)$ is open (i.e., it contains no boundary points), and it is differentiable at every point in $\mathcal{D}(f)$.

A differentiable function $f(\cdot)$ is convex if and only if $\mathcal{D}(f)$ is convex and

$$f(y) \geq f(x) + \nabla f(x)^T (y - x), \quad \forall x, y \in \mathcal{D}(f). \tag{4.7}$$

This inequality is called the *first-order condition* for convex functions.

Geometrically, the first-order condition (4.7) means that the line passing through $(x, f(x))$ along the gradient of $f(\cdot)$ at $x$, i.e., $\nabla f(x)$, lies under the graph of $f(\cdot)$. Figure 4.4 illustrates the first-order condition for a simple convex function on $\mathbb{R}$.

The first-order condition (4.7) is the most important property of convex functions, and plays an important role in convex optimization. It implies that from local information about a convex function (i.e., its value $f(x)$ and gradient $\nabla f(x)$ at a point $x$), we can derive global information (i.e., a global underestimator of $f(\cdot)$ at any point). As one simple example, the inequality (4.7) shows that if $\nabla f(x) = 0$ (i.e., $\nabla f(x)_i = 0$, $i = 1, ..., n$), then $x$ is a global minimizer of $f(\cdot)$, since $f(y) \geq f(x), \forall y \in \mathcal{D}(f)$.

**Second-order Conditions for Convex Functions**

The second derivative or *Hessian matrix* of a scalar-valued function $f(\cdot)$ at a point $\boldsymbol{x} \in \mathcal{D}(f)$, denoted by $\nabla^2 f(\boldsymbol{x})$, is an $n \times n$ matrix, given by

$$\nabla^2 f(\boldsymbol{x})_{ij} = \frac{\partial^2 f(\boldsymbol{x})}{\partial x_i \partial x_j}, \quad i = 1, ..., n, \, j = 1, ..., n, \tag{4.8}$$

provided that $f(\cdot)$ is twice differentiable at $\boldsymbol{x}$. We say $f(\cdot)$ is twice differentiable (everywhere in its domain) if $\mathcal{D}(f)$ is open, and it is twice differentiable at every point in $\mathcal{D}(f)$.

A twice differentiable $f(\cdot)$ is convex, if and only if $\mathcal{D}(f)$ is convex and

$$\nabla^2 f(\boldsymbol{x}) \succeq 0, \quad \forall \boldsymbol{x} \in \mathcal{D}(f), \tag{4.9}$$

that is, if its Hessian matrix is positive semidefinite. This inequality is referred to as the *second-order condition* for convex functions.

For a scalar-valued function $f(\cdot)$ on $\mathbb{R}$, the inequality (4.9) reduces to the simple condition $f''(x) \geq 0$, which means that the gradient is nondecreasing. Geometrically, the second-order condition (4.9) can be interpreted as the requirement that the graph of the function have positive (upward) curvature at $\boldsymbol{x}$.

**Operations Preserving Convexity of Functions**

Now we describe some simple operations that preserve convexity (or concavity) of functions, or allow us to construct new convex and concave functions.

1. *Nonnegative weighted sums:* Suppose $f_1(\cdot), ..., f_k(\cdot)$ are convex, and $\theta_1, ..., \theta_k \geq 0$. Define a new function by

$$f(\boldsymbol{x}) \triangleq \theta_1 f_1(\boldsymbol{x}) + ... + \theta_k f_k(\boldsymbol{x}),$$

with $\mathcal{D}(f) = \mathcal{D}(f_1) \cap ... \cap \mathcal{D}(f_k)$. Then $f(\cdot)$ is also convex.

2. *Composition with an affine mapping:* Suppose $f(\cdot)$ is a function on $\mathbb{R}^n$, $\mathbf{A} \in \mathbb{R}^{n \times m}$, and $\boldsymbol{b} \in \mathbb{R}^n$. Define a new function by

$$g(\boldsymbol{x}) \triangleq f(\mathbf{A}\boldsymbol{x} + \boldsymbol{b}),$$

with $\mathcal{D}(g) = \{\boldsymbol{x} \in \mathbb{R}^m | \mathbf{A}\boldsymbol{x} + \boldsymbol{b} \in \mathcal{D}(f)\}$. Then if $f(\cdot)$ is convex, so is $g(\cdot)$.

3. *Point-wise maximum:* Suppose $f_1(\cdot), ..., f_k(\cdot)$ are convex. Define a new function by their pointwise maximum

$$f(\boldsymbol{x}) \triangleq \max\{f_1(\boldsymbol{x}), ..., f_k(\boldsymbol{x})\},$$

with $\mathcal{D}(f) = \mathcal{D}(f_1) \cap ... \cap \mathcal{D}(f_k)$. Then $f(\cdot)$ is also convex.

## 4.1.2 CONVEX OPTIMIZATION

A *mathematical optimization problem* usually describes the problem of finding a point over a feasible set that minimizes an objective function. It has the form

$$
\begin{aligned}
\text{minimize} \quad & f(\boldsymbol{x}) \\
\text{subject to} \quad & f_i(\boldsymbol{x}) \le 0, \ i = 1, ..., m.
\end{aligned} \tag{4.10}
$$

The function $f : \mathbb{R}^n \to \mathbb{R}$ is called the *objective function* (or cost function), the functions $f_i : \mathbb{R}^n \to \mathbb{R}, i = 1, ..., m$, are called the (inequality) *constraint functions*, and the point $\boldsymbol{x} \in \mathbb{R}^n$ is the *optimization variable* of the problem.[2] The domain of an optimization problem (4.10) is the intersection of the objective function's domain and all constraint functions' domains, denoted by $\mathcal{D}(\mathrm{p})$ or $\mathcal{D}$ simply, i.e., $\mathcal{D} \triangleq \mathcal{D}(f) \cap \mathcal{D}(f_1) \cap ... \cap \mathcal{D}(f_m)$.

A point $\boldsymbol{x}$ is *feasible* for an optimization problem (4.10) if it satisfies all constraints $f_i(\boldsymbol{x}) \le 0, i = 1, ..., m$. The set of all feasible points is called the *constraint set* or *feasible set* for the optimization problem (4.10), denoted by $\mathcal{C}(\mathrm{p})$ or $\mathcal{C}$ simply, i.e.,

$$
\mathcal{C} \triangleq \{\boldsymbol{x} \mid \boldsymbol{x} \in \mathcal{D}, \ f_i(\boldsymbol{x}) \le 0, i = 1, ..., m\}. \tag{4.11}
$$

A feasible point $\boldsymbol{x} \in \mathcal{C}$ is called (*globally*) *optimal* (a global minimizer), or a solution of the problem (4.10), if it has the smallest objective value among all feasible points, i.e.,

$$
f(\boldsymbol{x}) \le f(\boldsymbol{z}), \ \forall \boldsymbol{z} \in \mathcal{C}. \tag{4.12}
$$

A feasible point $\boldsymbol{x} \in \mathcal{C}$ is *locally optimal* (a local minimizer), if it is no worse than its feasible neighbors, that is, if there is an $\epsilon > 0$ such that

$$
f(\boldsymbol{x}) \le f(\boldsymbol{z}), \ \forall \boldsymbol{z} \in \mathcal{C} \text{ with } ||\boldsymbol{z} - \boldsymbol{x}|| \le \epsilon, \tag{4.13}
$$

where $||\boldsymbol{x}|| \triangleq \sqrt{\boldsymbol{x}^T \boldsymbol{x}}$ is the standard Euclidean norm of a vector $\boldsymbol{x}$. That is, $||\boldsymbol{z} - \boldsymbol{x}||$ is the Euclidean distance between points $\boldsymbol{z}$ and $\boldsymbol{x}$.

We are interested in a class of optimization problems called *convex optimization problems*, where the objective and constraint functions are convex. This implies the domain $\mathcal{D}$ and the constraint set $\mathcal{C}$ are both convex. A fundamental property of convex optimization problem is that: *a local minimizer is also a global minimizer. If in addition the objective function is strictly convex, then the global minimizer is unique.*

### Unconstrained Convex Optimization

If there is no constraint (i.e., $m = 0$) in the problem (4.10), we say it is an *unconstrained* optimization. That is, an *unconstrained convex optimization* problem is one of the form

$$
\text{minimize} \quad f(\boldsymbol{x}) \tag{4.14}
$$

---

[2]Note that any equality constraint (e.g., $h(\boldsymbol{x}) = 0$) can be represented by two inequality constraints equivalently, i.e., $h(\boldsymbol{x}) \ge 0$ and $-h(\boldsymbol{x}) \ge 0$. Therefore, we consider the inequality constraint only without loss of generality.

where the objective function $f(\cdot)$ is convex. Obviously, in an unconstrained convex optimization, the constraint set is the domain of $f(\cdot)$, i.e., $\mathcal{C} = \mathcal{D}(f)$.

The following lemma characterizes the optimality conditions for an unconstrained convex optimization, that is, the necessary and sufficient conditions for a feasible point to be (globally) optimal.

**Lemma 4.5**    *Suppose the objective function $f(\cdot)$ is convex and differentiable. A feasible point $\boldsymbol{x}^* \in \mathcal{C}$ is a global minimizer of $f(\cdot)$ or a solution of (4.14) if and only if*

$$\nabla f(\boldsymbol{x}^*) = \boldsymbol{0}, \tag{4.15}$$

*that is,* $\nabla f(\boldsymbol{x}^*)_i = \frac{\partial f(\boldsymbol{x}^*)}{\partial x_i} = 0, i = 1, ..., n.$

We now discuss the computational methods for solving an unconstrained optimization. By the optimality condition (4.15), solving the problem (4.14) is the same as finding a solution of $\nabla f(\boldsymbol{x}) = \boldsymbol{0}$, i.e., a set of $n$ equations $\frac{\partial f(\boldsymbol{x})}{\partial x_i} = 0, i = 1, ..., n$. In a few special cases, it is possible to solve these $n$ equations analytically; but more generally the problem must be solved by an iterative algorithm. That is, we want to find an algorithm that computes a sequence of feasible points $\boldsymbol{x}^{(0)}, \boldsymbol{x}^{(1)}, ...$ with $f(\boldsymbol{x}^{(k)}) \to f(\boldsymbol{x}^*)$ as $k \to \infty$. Such a sequence of points is called a *minimizing sequence* for the problem (4.14). An algorithm is said to be *iteratively descent*, if it successively generates points $\boldsymbol{x}^{(1)}, \boldsymbol{x}^{(2)}, ...$, (from an initial point $\boldsymbol{x}^{(0)}$) such that $f(\cdot)$ is decreasing at each iteration, i.e., $f(\boldsymbol{x}^{(k+1)}) < f(\boldsymbol{x}^{(k)}), \forall k$.

We focus on the most popular gradient-based algorithms, which have the form

$$\boldsymbol{x}^{(k+1)} = \boldsymbol{x}^{(k)} + \gamma^{(k)} \boldsymbol{d}^{(k)},$$

where $\gamma^{(k)} > 0$ is a positive scalar called the *step size* or *step length* at iteration $k$, and $\boldsymbol{d}^{(k)}$ is a gradient-based $n$-vector called the *step* or *search direction* at iteration $k$. The iterative descent property requires that $\nabla f(\boldsymbol{x}^{(k)})^T \boldsymbol{d}^{(k)} < 0$, otherwise, $f(\boldsymbol{x}^{(k+1)}) - f(\boldsymbol{x}^{(k)}) \geq \gamma^{(k)} \nabla f(\boldsymbol{x}^{(k)})^T \boldsymbol{d}^{(k)} \geq 0$. Two widely used gradient-based algorithms are presented below. For more algorithms, please refer to [20].

1. *Gradient Descent Method:*

$$\boldsymbol{d}^{(k)} \triangleq -\nabla f(\boldsymbol{x}^{(k)}).$$

That is, the search direction at each iteration $k$ is the negative gradient at $\boldsymbol{x}^{(k)}$.

2. *Newton's Method:*

$$\boldsymbol{d}^{(k)} \triangleq -\left(\nabla^2 f(\boldsymbol{x}^{(k)})\right)^{-1} \nabla f(\boldsymbol{x}^{(k)}).$$

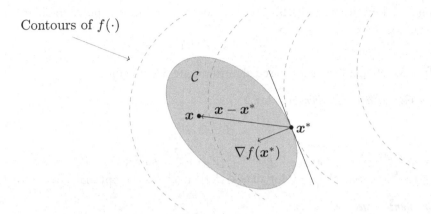

Contours of $f(\cdot)$

**Figure 4.5:** Geometric interpretation of the optimality condition in (4.17). The constraint set $\mathcal{C}$ is shown shaded. The gradient $\nabla f(x^*)$ makes an angle less than or equal to 90 degrees with all feasible variations $x - x^*$.

### Constrained Convex Optimization

A *constrained convex optimization* problem, or just convex optimization, is one of the form

$$
\begin{aligned}
&\text{minimize} \quad f(x) \\
&\text{subject to} \quad f_i(x) \leq 0, \ i = 1, ..., m,
\end{aligned}
\tag{4.16}
$$

where the objective function $f(\cdot)$ and the constraint functions $f_i(\cdot)$ are convex. According to (4.11), the constraint set $\mathcal{C}$ is also convex.

The following lemma characterizes the optimality conditions for a constrained convex optimization, that is, the necessary and sufficient conditions for a feasible point to be (globally) optimal.

**Lemma 4.6** *Suppose the objective function $f(\cdot)$ is convex and differentiable. A feasible point $x^* \in \mathcal{C}$ is a global minimizer of $f(\cdot)$ or a solution of (4.16) if and only if*

$$
\nabla f(x^*)^T (x - x^*) \geq 0, \quad \forall x \in \mathcal{C}.
\tag{4.17}
$$

To better understand the optimality condition (4.17), we illustrate it geometrically in Figure 4.5. At a minimizer $x^*$, the gradient $\nabla f(x^*)$ makes an angle less than or equal to 90 degrees with all feasible variations $x - x^*$, so that $\nabla f(x^*)^T (x - x^*) \geq 0, \forall x \in \mathcal{C}$.

We now turn to the computational methods for solving a constrained optimization problem. Although there is a great variety of algorithms for this problem, we will restrict ourselves to a limited class of methods that generate a minimizing sequence of feasible $x^{(k)}$ by searching along descent directions (i.e., iterative descent). Similarly, we focus on the most popular gradient-based algorithms, which have the form

$$x^{(k+1)} = x^{(k)} + \gamma^{(k)} d^{(k)}.$$

Two widely used gradient-based algorithms are listed below.

1. *The Conditional Gradient Method:*

$$d^{(k)} \triangleq \overline{x}^{(k)} - x^{(k)},$$

where $\overline{x}^{(k)} \triangleq \arg\max_{x \in C} \nabla f(x^{(k)})^T (x - x^{(k)})$ subject to $\nabla f(x^{(k)})^T (x - x^{(k)}) < 0$. If $\nabla f(x^{(k)})^T (x - x^{(k)}) \geq 0$ for all $x \in C$, then $x^{(k)}$ is the optimal solution by (4.17).

2. *Gradient Projection Method:*

$$d^{(k)} \triangleq \overline{x}^{(k)} - x^{(k)},$$

where $\overline{x}^{(k)}$ is given by $\overline{x}^{(k)} \triangleq \left[ x^{(k)} - s^{(k)} \nabla f(x^{(k)}) \right]^+$. Here $[\cdot]^+$ denotes projection on the constraint set $C$, and $s^{(k)}$ is a positive scalar.

## 4.1.3   DUALITY PRINCIPLE

Now we introduce the *Lagrangian duality*, which plays a central role in convex optimization. By duality principle, an optimization problem (which we refer to as the *primal problem*) can usually be converted into a (Lagrange) dual form, which is termed a (Lagrange) *dual problem*. The solution of the dual problem provides a lower bound to the solution of the primal problem. In addition if the primal problem is convex and satisfies a constraint qualification, then the value of an optimal solution of the primal problem is given by the dual problem [19].

**Lagrange Dual Functions**

The basic idea in Lagrangian duality is to take the constraints into account by adding the objective function with a weighted sum of the constraint functions. The weight associated with each constraint function $f_i(x)$ is referred to as the *Lagrange multiplier*, denoted by $\lambda_i$. The vector $\lambda \triangleq (\lambda_1, ..., \lambda_m)$ is called the *dual variable* or *Lagrange multiplier vector*. The *Lagrangian function* and *dual function* for problem (4.16) are defined as follows.[3]

**Definition 4.7   Lagrangian Function**   The *Lagrangian function* (or just *Lagrangian*) $L : \mathbb{R}^n \times \mathbb{R}^m \to \mathbb{R}$ is defined as

$$L(x, \lambda) \triangleq f(x) + \sum_{i=1}^{m} \lambda_i f_i(x), \tag{4.18}$$

with the domain $\mathcal{D}(L) = \mathcal{D} \times \mathbb{R}^m$, where $\lambda_i \geq 0$, and $\mathcal{D} = \mathcal{D}(f) \cap \mathcal{D}(f_1) \cap ... \cap \mathcal{D}(f_m)$ is the domain of the optimization problem (4.16).

---

[3]Note the following discussions are not only applicable to convex optimization problems, but also to non-convex optimization problems.

**Definition 4.8** **Dual Function** The (*Lagrange*) *dual function* $g : \mathbb{R}^m \to \mathbb{R}$ is defined as the minimum value of the Lagrangian over $\boldsymbol{x}$:

$$g(\boldsymbol{\lambda}) \triangleq \inf_{\boldsymbol{x} \in \mathcal{D}} L(\boldsymbol{x}, \boldsymbol{\lambda}) = \inf_{\boldsymbol{x} \in \mathcal{D}} \left( f(\boldsymbol{x}) + \sum_{i=1}^{m} \lambda_i f_i(\boldsymbol{x}) \right). \tag{4.19}$$

The dual function $g(\cdot)$ is always concave (even when the original problem (4.16) is not convex), since it is the pointwise infimum of a family of affine functions of $\boldsymbol{\lambda}$.

One of the most important properties for the dual function $g(\cdot)$ is that it yields lower bounds on the optimal value $f(\boldsymbol{x}^*)$ of the problem (4.16): for any $\boldsymbol{\lambda} \succeq \boldsymbol{0}$ we have

$$g(\boldsymbol{\lambda}) = \inf_{\boldsymbol{x} \in \mathcal{D}} L(\boldsymbol{x}, \boldsymbol{\lambda}) \leq \inf_{\boldsymbol{x} \in \mathcal{C}} L(\boldsymbol{x}, \boldsymbol{\lambda}) \leq L(\boldsymbol{x}^*, \boldsymbol{\lambda}) \leq f(\boldsymbol{x}^*). \tag{4.20}$$

The first inequality follows because the constraint set is a subset of the domain, i.e., $\mathcal{C} \subseteq \mathcal{D}$, the second inequality follows because the optimal point $\boldsymbol{x}^*$ lies in the constraint set, i.e., $\boldsymbol{x}^* \in \mathcal{C}$, and the last inequality follows because $f_i(\boldsymbol{x}) \leq 0$ for any feasible $\boldsymbol{x} \in \mathcal{C}$.

**Lagrange Dual Problems**
As shown in (4.20), the Lagrange dual function $g(\cdot)$ yields a lower bound on the optimal value $f(\boldsymbol{x}^*)$ of the optimization problem (4.16), and how far the dual function $g(\cdot)$ is apart from the optimal value $f(\boldsymbol{x}^*)$ essentially depends on the dual variable $\boldsymbol{\lambda}$. Thus, a natural question is: What is the best lower bound that can be obtained from the Lagrange dual function? This leads to the following optimization problem

$$\begin{aligned} \text{maximize} \quad & g(\boldsymbol{\lambda}) \\ \text{subject to} \quad & \boldsymbol{\lambda} \succeq \boldsymbol{0}. \end{aligned} \tag{4.21}$$

The problem (4.21) is called the (*Lagrange*) *dual problem* associated with the problem (4.16), which we call the *primal problem* in this context. Obviously, the dual problem (4.21) is convex (even when the primal problem (4.16) is not convex), since the objective to be maximized is concave and the constraint set is convex. Therefore, the solution of (4.21) is given by Lemma 4.6 (suppose the dual function $g(\cdot)$ is differentiable).

Let $\boldsymbol{\lambda}^*$ denote a solution (a global maximizer) of the dual problem (4.21). For clarity, we refer to $\boldsymbol{\lambda}^*$ as *dual optimal* or *optimal Lagrange multipliers*, and $\boldsymbol{x}^*$, a solution of the primal problem (4.16), as *primal optimal*. The optimal value $g(\boldsymbol{\lambda}^*)$ of the dual problem (4.21), by definition, is the best lower bound on $f(\boldsymbol{x}^*)$ that can be obtained from the dual function. In particular, we have

$$g(\boldsymbol{\lambda}^*) \leq f(\boldsymbol{x}^*). \tag{4.22}$$

This property is called the *weak duality*. The difference $f(\boldsymbol{x}^*) - g(\boldsymbol{\lambda}^*)$ is called the *optimal duality gap* between the primal problem and the dual problem, which is always nonnegative.

If the optimal duality gap attains zero, that is,

$$g(\boldsymbol{\lambda}^*) = f(\boldsymbol{x}^*), \tag{4.23}$$

then we say that *strong duality* holds.

However, strong duality does not always hold, even when the primal problem is convex. There are a lot of results that establish conditions on the problem under which strong duality holds. These conditions are called *constraint qualifications* [21].

**KKT Optimality Conditions**

Now suppose strong duality holds. For any feasible point $\boldsymbol{x}$ of the primal problem (4.16) and $\boldsymbol{\lambda}$ of the dual problem (4.21), we have

$$f(\boldsymbol{x}) - f(\boldsymbol{x}^*) \le f(\boldsymbol{x}) - g(\boldsymbol{\lambda}), \tag{4.24}$$

since $g(\boldsymbol{\lambda}) \le f(\boldsymbol{x}^*)$. Hence, dual feasible points allow us to bound how suboptimal a given feasible point is, without knowing the exact value of $f(\boldsymbol{x}^*)$.

We refer to the gap between primal and dual objectives, i.e., $f(\boldsymbol{x}) - g(\boldsymbol{\lambda})$, as the *duality gap* associated with the primal feasible point $\boldsymbol{x}$ and dual feasible point $\boldsymbol{\lambda}$. Any primal-dual feasible pair $\{\boldsymbol{x}, \boldsymbol{\lambda}\}$ localizes the optimal values of the primal and dual problems to an interval $[g(\boldsymbol{\lambda}), f(\boldsymbol{x})]$, that is,

$$g(\boldsymbol{\lambda}) \le g(\boldsymbol{\lambda}^*) \le f(\boldsymbol{x}^*) \le f(\boldsymbol{x}). \tag{4.25}$$

Obviously, if the duality gap of a primal-dual feasible pair $\{\boldsymbol{x}, \boldsymbol{\lambda}\}$ is zero, i.e., $g(\boldsymbol{\lambda}) = f(\boldsymbol{x})$, then $\boldsymbol{x}$ is the primal optimal, $\boldsymbol{\lambda}$ is the dual optimal, and strong duality holds.

Let $\boldsymbol{x}^*$ be a primal optimum and $\boldsymbol{\lambda}^*$ be a dual optimum. By strong duality, we have

$$f(\boldsymbol{x}^*) = g(\boldsymbol{\lambda}^*) = \inf_{\boldsymbol{x} \in \mathcal{D}} L(\boldsymbol{x}, \boldsymbol{\lambda}^*) \le L(\boldsymbol{x}^*, \boldsymbol{\lambda}^*) \le f(\boldsymbol{x}^*). \tag{4.26}$$

The first equality states that the optimal duality gap is zero, the second equality follows the definition of the dual function, the third inequality follows because the primal optimal $\boldsymbol{x}^* \in \mathcal{C} \subseteq \mathcal{D}$, and the last inequality follows because $\boldsymbol{\lambda} \succeq \mathbf{0}$ and $f_i(\boldsymbol{x}) \le 0, i = 1, ..., m$.

We can draw several interesting conclusions from (4.26). Firstly, the last inequality is indeed an equality, which implies that $\sum_{i=1}^{m} \lambda_i^* f_i(\boldsymbol{x}^*) = 0$; since each term in the sum is non-positive, we further have $\lambda_i^* f_i(\boldsymbol{x}^*) = 0, i = 1, ..., m$. This condition is referred to as the *complementary slackness*, which holds for any primal optimal $\boldsymbol{x}^*$ and any dual optimal $\boldsymbol{\lambda}^*$ (when strong duality holds). The complementary slackness can also be expressed as

$$\lambda_i^* > 0 \Rightarrow f_i(\boldsymbol{x}^*) = 0,$$

or, in other words,

$$f_i(\boldsymbol{x}^*) > 0 \Rightarrow \lambda_i^* = 0.$$

Roughly speaking, this means the $i$-th optimal Lagrange multiplier is zero unless the $i$-th constraint is active at the optimum.

Secondly, the third inequality is also an equality, i.e., $\inf_{x \in \mathcal{D}} L(x, \lambda^*) = L(x^*, \lambda^*)$, which implies that $x^*$ minimizes the Lagrangian $L(x, \lambda^*)$. This means that

$$\frac{\partial L(x^*, \lambda^*)}{\partial x} = \nabla f(x^*) + \sum_{i=1}^{m} \lambda_i^* \nabla f_i(x^*) = 0. \tag{4.27}$$

Based on above, we can obtain the necessary and sufficient conditions for a primal dual feasible pair $\{x^*, \lambda^*\}$ to be optimal (for the primal problem and dual problem, respectively). We refer to these conditions as the *Karush-Kuhn-Tucker (KKT)* conditions [22].

**Lemma 4.9  Karush-Kuhn-Tucker (KKT) Conditions**  *Assume that the primal problem is strictly convex and the strong duality holds. A primal dual feasible pair $\{x^*, \lambda^*\}$ is optimal for the primal problem and dual problem, respectively, if and only if*

$$\begin{cases} f_i(x^*) \leq 0, \ \lambda_i^* \geq 0, \ \lambda_i^* \cdot f_i(x^*) = 0, \quad i = 1, ..., m \\ \nabla f(x^*) + \sum_{i=1}^{m} \lambda_i^* \nabla f_i(x^*) = 0. \end{cases} \tag{4.28}$$

According to Lemma 4.9, solving the primal problem (4.16) is the same as finding the primal dual feasible pairs $\{x, \lambda\}$ that satisfy the KKT conditions in (4.28). In a few special cases it is possible to solve the KKT conditions (and therefore, the optimization problem) analytically. Generally, the KKT conditions are solved by an iterative algorithm.

**Shadow Price**

There are some interesting interpretations for the Lagrange multipliers $\lambda_i$, $i = 1, ..., m$. Now we can give a simple geometric interpretation of the Lagrange multipliers in terms of economics, where they are often interpreted as *prices*.

To show this, we first introduce the perturbed version of the original problem (4.16)

$$\begin{aligned} \text{minimize} \quad & f(x) \\ \text{subject to} \quad & f_i(x) \leq u_i, \ i = 1, ..., m, \end{aligned} \tag{4.29}$$

where $u_i$ is the perturbing parameter for the $i$-th inequality constraint. That is, when $u_i$ is positive, it means that we have relaxed the $i$-th constraint; when $u_i$ is negative, it means that we have tightened the constraint. This perturbed problem coincides with the original problem (4.16) when $u \triangleq (u_1, ..., u_m) = 0$.

The optimal value of the perturbed problem (4.29) is given by

$$p^*(u) \triangleq \inf_{x \in \mathcal{C}(u)} f(x),$$

where $C(u) \triangleq \{x \mid f_i(x) \leq u_i, i = 1, ..., m\}$ is the constraint set of the perturbed problem (4.29). Note that both the constraint set $C(u)$ and the optimal value $p^*(u)$ of the perturbed problem (4.29) depend on the perturbing parameters $u_i, i = 1, ..., m$. When $u = 0$, we have $C(u) = C$ where $C$ is the constraint set of the original problem (4.16), and $p^*(0) = f(x^*)$ where $f(x^*) \triangleq \inf_{x \in C} f(x)$ is the optimal value of the original problem (4.16).

Suppose $x \in C(u)$ is any feasible point for the perturbed problem (4.29), i.e., $f_i(x) \leq u_i$, $i = 1, ..., m$. For any perturbing parameters $u$ and feasible point $x \in C(u)$, we have

$$p^*(0) = f(x^*) = g(\lambda^*) \leq f(x) + \sum_{i=1}^{m} \lambda_i^* f_i(x) \leq f(x) + \sum_{i=1}^{m} \lambda_i^* u_i.$$

The second equality follows from the strong duality, the third inequality follows from the definition of $g(\lambda^*)$, and the last inequality follows because $f_i(x) \leq u_i$ and $\lambda_i^* \geq 0$, $i = 1, ..., m$. Since the above formula holds for any feasible point $x \in C(u)$, we have

$$p^*(u) \triangleq \inf_{x \in C(u)} f(x) \geq p^*(0) - \sum_{i=1}^{m} \lambda_i^* u_i.$$

Suppose now that $p^*(u)$ is differentiable at $u = 0$. Then we have

$$\frac{\partial p^*(0)}{\partial u_i} = -\lambda_i^*.$$

Now we give a simple interpretation of the above result in terms of economics. As we view the variable $x$ as a firm's investments on $n$ different resources, the objective $f(\cdot)$ as the firm's cost, or $-f(\cdot)$ as the firm's profit, and each constraint $f_i(x) \leq 0$ as a limit on some resource investments. The (negative) perturbed optimal cost function $-p^*(u)$ tells us how much more or less profit could be made if more, or less, of each resource were made available to the firm. In other words, when $u$ is close to $0$, the Lagrange multiplier $\lambda_i^*$ tells us approximately how much more profit the firm could make, for a small increase in the availability of resource $i$. Thus, $\lambda_i^*$ can be viewed as the natural or equilibrium *price* for resource $i$. For this reason a dual optimal $\lambda^*$ is sometimes called *shadow prices*.

### Solving Dual Problem Using the Subgradient Method

We now consider the methods for solving dual problems, also called dual methods. There are several incentives for solving the dual problem in place of the primal: (a) The dual problem is a convex optimization, while the primal may not be convex; (b) The dual problem may have smaller dimension and/or simpler constraints than the primal; (c) If strong duality holds, the dual optimal value is exactly the primal optimal value; and (d) Even if strong duality does not hold, the dual optimal value provides a lower bound to the primal optimal value, which may be useful in designing iterative algorithms.

Of course, we should also consider some of the difficulties in solving the dual problem. The most critical ones are the following: (a) The evaluation of the dual function $g(\lambda)$ at any dual variable

$\lambda$ requires minimization of the Lagrangian $L(x, \lambda)$ over all $x \in \mathcal{D}$; (b) The dual function $g(\lambda)$ may not be differentiable in many types of problems; and (c) If strong duality does not hold, there is a certain duality gap between the dual optimal and primal optimal.

We will discuss an important type of dual method, namely the *subgradient method*, which is particularly suitable for solving a dual problem with nondifferentiable objective function. The basic idea of subgradient methods is to generate a minimizing sequence of dual feasible $\lambda^{(k)}$ using subgradients rather than gradients as search direction.

Given a convex function $f : \mathbb{R}^n \to \mathbb{R}$, we say that a vector $d \in \mathbb{R}^n$ is a *subgradient* of $f(\cdot)$ at a feasible point $x \in \mathcal{D}(f)$ if

$$f(z) \geq f(x) + d^T(z - x), \quad \forall z \in \mathcal{D}(f). \tag{4.30}$$

If instead $f(\cdot)$ is a concave function, we say that $d$ is a subgradient of $f(\cdot)$ at point $x$ if and only if $-d$ is a subgradient of $-f(\cdot)$ at $x$. This means that a subgradient $d$ of the dual function $g(\lambda)$ at a dual feasible point $\lambda \in \mathcal{D}(g)$ satisfies:

$$g(\mu) \leq g(\lambda) + d^T(\mu - \lambda), \quad \forall \mu \in \mathcal{D}(g). \tag{4.31}$$

Thus, the subgradient method generates a minimizing sequence of dual feasible $\lambda^{(k)}$ according to the following iteration

$$\lambda^{(k+1)} = \left[\lambda^{(k)} + \gamma^{(k)} d^{(k)}\right]^+, \tag{4.32}$$

where $\gamma^{(k)}$ is the step-size, $d^{(k)}$ is the subgradient of $g(\lambda)$ at point $\lambda^{(k)}$, and $[\lambda]^+$ denotes the projection of $\lambda$ on the constraint set of the dual problem (4.21).

The fundamental difference between the gradient-based method and the subgradient method is that with the subgradient method, the new iterative may not improve the dual objective for all values of the step-size $\gamma^{(k)}$. That is, for some large $\gamma^{(k)}$ we may have $g(\lambda^{(k+1)}) < g(\lambda^{(k)})$, whereas for sufficiently small step-size $\gamma^{(k)}$ we have $g(\lambda^{(k+1)}) \geq g(\lambda^{(k)})$. This is shown in the following Lemma, which also provides an estimate for the range of appropriate step-sizes.

**Lemma 4.10**   *For every dual optimal solution $\lambda^*$, we have $||\lambda^{(k+1)} - \lambda^*|| < ||\lambda^{(k)} - \lambda^*||$ for all step-sizes $\gamma^{(k)}$ satisfying*

$$0 < \gamma^{(k)} < 2 \cdot \frac{g(\lambda^*) - g(\lambda^{(k)})}{||d^{(k)}||^2}. \tag{4.33}$$

Unfortunately, the above range for $\gamma^{(k)}$ requires the dual optimal value $g(\lambda^*)$, which is usually unknown. In practice, we can use the following approximate step-size formula

$$\gamma^{(k)} = \alpha^{(k)} \cdot \frac{g^{(k)} - g(\lambda^{(k)})}{||d^{(k)}||^2}, \tag{4.34}$$

where $g^{(k)}$ is an approximation to the optimal dual value and $0 < \alpha^{(k)} < 2$.

There are many variations of subgradient methods that aim to accelerate the convergence of the basic method. For more details, please refer to [19, 20].

## 4.2   APPLICATION I: RESOURCE ALLOCATION FOR WIRELESS VIDEO STREAMING

With the advances of mobile computing technologies and deployments of new cellular infrastructure, video communications are becoming more important in many new business applications. However, there are still many open problems in terms of how to efficiently provision complicated video QoS requirements for mobile users. One particularly challenging problem is multi-user video streaming over wireless channels, where the demand for better video quality and small transmission delays needs to be reconciled with the limited and often time-varying communication resources. We will resolve this issue through a new framework for resource allocation, source adaptation, and deadline oriented scheduling.

### 4.2.1   NETWORK OPTIMIZATION FRAMEWORK

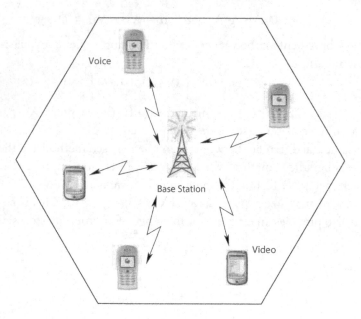

**Figure 4.6:**  A single cell network with mixed voice and video users.

We consider a single cell in a CDMA-based cellular network, with a mixed population of voice and video users, as shown in Fig. 4.6. A voice transmission is successful if a target Signal-to-Interference-plus-Noise Ratio (SINR) is reached at the receiver. A video user is more flexible and can adapt to the network environment in terms of the achieved SINR and the transmission rate. However, once the video frames are transmitted, stringent delay deadlines need to be satisfied to guarantee the streaming quality.

The network's goal is to maximize the overall quality of the video users, subject to the QoS constraints of voice users. To achieve this, we need to carefully optimize network resource allocation, video source signal processing, and video frame scheduling.

We will consider both uplink and downlink video streaming. In the uplink case where users stream videos to the base station, video users need to limit aggregate interference that they generate and affect the voice users. In the downlink case where users stream videos frame the base station, the network needs to limit the amount of transmission power allocated to the video users. In both cases, the optimal video streaming problem can be modeled in the framework of nonlinear constrained optimization. Two key questions that need to be answered are: (i) how to allocate resources among video users in an efficient manner (i.e., maximizing total users' quality or minimizing total users' distortion)? (ii) how to make sure that the stringent delivery deadline requirements are met for every video frame that is chosen for transmission?

The general solution framework that answers the above two questions involves three phases:

1. *Average resource allocation.* This can be formulated as a constrained network optimization problem, which exploits the multiuser content diversity to efficiently utilize the network resources.

2. *Video source adaptations.* Based on the average resource allocation results in phase 1, each video user adapts the video source by solving a localized optimization problem with video summarization.

3. *Multiuser deadline oriented scheduling.* The network decides a transmission schedule based on video users' source decisions in phase 2, in order to meet the stringent deadline constraints of the streaming applications.

In some cases we may not be able to find a feasible scheduling in phase 3. This implies that although the system resource is enough in an average sense (guaranteed by phase 1), the deadline requirements might be too stringent to satisfy. In that case, we will go back to phase 1 and re-optimize the average resource allocation, but with more stringent resource constraints. This will force the users to be more conservative when doing the source adaptations in phase 2 (i.e., each video user will transmit fewer frames), thus make it easier to achieve a feasible scheduling in phase 3. Here we will focus the discussions on phase 1, and leave the discussions of phases 2 and 3 to [23].

## 4.2.2 AVERAGE RESOURCE ALLOCATION

Assume there are $N$ video users in the cell. We characterize the QoS of a video user $n$ by a utility function $u_n(x_n)$, which is an increasing and strictly concave function of the communication resource $x_n$ allocated to user $n$. This models various commonly used video quality measures such as the rate-PSNR function and rate-summarization distortion functions. It is well known from information theory [24] that the rate-distortion functions for a variety of sources are convex, and in practice, the operational rate distortion functions are usually convex as well. Thus, the utility functions (defined as negative distortion) are concave. For the average resource allocation phase, we assume that $u_n(x_n)$ is continuous in $x_n$.

The average resource allocation is achieved by solving the following optimization problem, where $X_{\max}$ denotes the total limited resource available to the video users (i.e., total transmission power in the downlink case and total transmission time in the uplink case),

$$\max_{\{x_n \geq 0, 1 \leq n \leq N\}} \sum_n u_n (x_n), \text{ s.t. } \sum_n x_n \leq X_{\max}. \tag{4.35}$$

Next we will solve Problem (4.35) using the dual decomposition technique introduced in Section 4.1.3, where the base station sets a price on the resource, and each mobile user determines its average resource request depending on the price and its own utility function in a distributed fashion.

More specifically, the dual-based decomposition works as follows. First, we relax the constraint in (4.35) with a dual variable $\lambda$ and obtain the following Lagrangian

$$L(\boldsymbol{x}, \lambda) \triangleq \sum_n u_n (x_n) - \lambda \left( \sum_n x_n - X_{\max} \right), \tag{4.36}$$

where $\boldsymbol{x} = (x_n, 1 \leq n \leq N)$. Then Problem (4.35) can be solved at two levels.

- At the lower level, each video user solves the following problem,

$$\max_{x_n \geq 0} (u_n (x_n) - \lambda x_n), \tag{4.37}$$

which corresponds to maximizing the surplus (i.e., utility minus payment) based on price $\lambda$. Denote the optimal solution of (4.37) as $x_n (\lambda)$, which is unique since the utility function is continuous, increasing, and strictly concave. The video users then feedback the values of $x_n (\lambda)$ to the base station.

- At the higher level, the base station adjusts $\lambda$ to solve the following problem

$$\min_{\lambda \geq 0} g(\lambda) \triangleq \sum_n g_n(\lambda) + \lambda X_{\max}, \tag{4.38}$$

where $g_n(\lambda)$ is the maximum value of (4.37) for a given value of $\lambda$. The dual function $g(\lambda)$ is non-differentiable in general, and (4.38) can be solved using the sub-gradient searching method,

$$\lambda^{(k+1)} = \max \left\{ 0, \lambda^{(k)} + \alpha^{(k)} \left( \sum_n x_n \left( \lambda^{(k)} \right) - X_{\max} \right) \right\}, \tag{4.39}$$

where $k$ is the iteration index and $\alpha^{(k)}$ is a small step size at iteration $k$.

Given the assumption on the utility functions, we have the property of strong duality which implies zero duality gap. In other words, given the optimal dual solution $\lambda^*$, the corresponding $x_n(\lambda^*)$ for all $n$ are the optimal solution of the primal problem (4.35).

The complete distributed algorithm is given in Algorithm 1, which converges under properly chosen small step sizes [25].

Next we give two concrete examples of Problem (4.37) for wireless uplink and downlink streaming.

---

**Algorithm 1** Dual-based Optimization Algorithm to solve Problem (4.35)

---

1: Initialization: set iteration index $k = 0$, and choose $0 < \epsilon \ll 1$ as the stopping criterion.

2: Base station announces an initial price $\lambda^{(k)} > 0$.

3: **repeat**

4:    **for all** video user $n$ **do**

5:        Compute $x_n\left(\lambda^{(k)}\right) = \arg\max_{x_n \geq 0}\left(u_n\left(x_n\right) - \lambda^{(k)} x_n\right)$.

6:        Send the value of $x_n\left(\lambda^{(k)}\right)$ to the base station.

7:    **end for**

8:    Base station updates the price $\lambda^{(k+1)} = \max\left\{0, \lambda^{(k)} + \alpha^{(k)}\left(\sum_n x_n\left(\lambda^{(k)}\right) - X_{\max}\right)\right\}$.

9:    $k = k + 1$.

10: **until** $|\lambda^{(k)} - \lambda^{(k-1)}| < \epsilon$.

---

## 4.2.3   WIRELESS UPLINK STREAMING

In a wireless CDMA network, different users transmit using different spreading codes. These codes are mathematically orthogonal under synchronous reception. However, the orthogonality is partially destroyed when the transmissions are asynchronous, such as in the uplink transmissions. The received SINR in that case is determined by the users' transmission power, the spreading factors (defined as the ratio of the bandwidth and the achieved rate), the modulation scheme used, and the background noise. The maximum constrained resource of the video users can be expressed as the maximum received power at the base station, derived based on a physical layer model similar as the one used in [26].

We consider the uplink transmission in a single CDMA cell with $M$ voice users and $N$ video streaming users. The total bandwidth $W$ is fixed and shared by all users. Each voice user has a QoS requirement represented in bit error rates (BER) (or frame error rates (FER)), which can be translated into a target SINR at the base station, $\gamma_{voice}$. Each voice user also has a target rate constraint $R_{voice}$. Assuming perfect power control, each voice user achieves the same received power at the base station, $P_{voice}^r$. The total received power at the base station from all video users is denoted as $P_{video}^{r,all}$. The background noise $n_0$ is fixed and includes both thermal noise and inter-cell interferences.

In order to support the successful transmissions of all voice users, we need to satisfy

$$\frac{W}{R_{voice}} \frac{G_{voice} P_{voice}^r}{n_0 W + (M-1) P_{voice}^r + P_{video}^{r,all}} \geq \gamma_{voice}. \tag{4.40}$$

Here $W/R_{voice}$ is the spreading factor, and coefficient $G_{voice}$ reflects the fixed modulation and coding schemes used by all voice users (e.g., $G_{voice} = 1$ for BPSK and $G_{voice} = 2$ for QPSK). For each voice user, the received interference comes from the other $M - 1$ voice users and all video users. From (4.40), we can solve for the maximum allowed value of $P_{video}^{r,all}$, denoted as

$$P_{video}^{r,\max} = \left(\frac{W G_{voice}}{R_{voice} \gamma_{voice}} - (M-1)\right) P_{voice}^r - n_0 W, \tag{4.41}$$

which is assumed to be fixed given fixed number of voice users $M$.

The network objective is to choose the transmission power of each video user during a time segment $[0, T]$, such that the total video's utility is maximized, i.e.,

$$\max_{\{p_n(t), \forall n\}} \sum_{n=1}^{N} u_n \left( \int_0^T r_n(\boldsymbol{p}(t)) \, dt \right) \tag{4.42}$$

$$\text{s.t.} \sum_{n=1}^{N} h_n p_n(t) \leq P_{video}^{r,\max}, \forall t \in [0, T]$$

$$0 \leq p_n(t) \leq P_n^{\max}, \forall n,$$

where $p_n(t)$ is the transmission power of video user $n$ at time $t$, $\boldsymbol{p}(t) = (p_n(t), \forall n)$ is the vector of all video users' transmission power at time $t$, $P_n^{\max}$ is the maximum peak transmission power of user $n$, and $h_n$ is the fixed channel gain from the transmitter of user $n$ to the base station. $r_n(t)$ is the rate achieved by user $n$ at time $t$, and depends on all video users' transmission power, the channel gains, the background noise, and interference from voice users. A user $n$'s utility function $u_n$ is defined on the video summarization quality of its transmitted sequence during $[0, T]$.

Problem (4.42) is not a special case of Problem (4.35), since (i) Problem (4.42) optimizes over $N$ functions ($p_n(t), \forall n$), whereas Problem (4.35) optimizes over $N$ variables ($x_n, \forall n$), and (ii) the objective function in Problem (4.42) is coupled across users, whereas the objective in Problem (4.35) is fully decoupled. This makes (4.42) difficult to solve in a distributed fashion.

In order to solve Problem (4.42), we will resort to the framework described in Section 4.2.1, where we will perform average resource allocation (in terms of average transmission power), source adaptation (to match the average resource allocation), and the deadline scheduling (to determine the exact power allocation functions by deadline aware water-filling).

To simplify the problem, let us assume that video users transmit in a TDMA fashion (but still concurrently with voice users). This is motivated by [27], where the authors showed that in order to achieve maximum total rate in a CDMA uplink, it is better to transmit weak power users in groups and strong power users one by one. Since video users typically need to achieve much higher rate than voice users (thus transmit at much higher power), it is reasonable to avoid simultaneous transmissions among video users, and thus avoid large mutual interference. A more important motivation for TDM transmission here is to exploit the temporal variation of the video contents, i.e., content diversity. Under such a TDM transmission scheme, the constrained resource to be allocated to the video users becomes the total transmission time of length $T$. The total number of bits that can be transmitted by user $n$ is determined by the transmission time allocated to it, $t_n \in [0, T]$, and the maximum rate it can achieve while it is allowed to transmit. Let us denote this rate as $R_n^{TDM}$, and it can be calculated by,

$$R_n^{TDM} = W \log_2 \left( 1 + \frac{\min\{h_n P_n^{\max}, P_{video}^{r,\max}\}}{n_0 W + M P_{voice}^r} \right). \tag{4.43}$$

Under the assumption of TDM transmission, Problem (4.42) can be written as

$$\max_{\{t_n \geq 0, \forall n\}} \sum_{n=1}^{N} u_n \left( R_n^{TDM} t_n \right), \text{ s.t. } \sum_{n=1}^{N} t_n \leq T. \tag{4.44}$$

Problem (4.44) is a special case of Problem (4.35), and the optimal transmission time allocation per user can be found using Algorithm 1.

Once the transmission time allocations are determined, each user locally adapts its source using summarization, which leads to the best sequence of video frames that fit into the transmission time allocation $t_n$. The transmission of each frame needs to meet a certain delivery deadline, after which the frame becomes useless. This requires the base station to determine a transmission schedule for all users, as shown in Section III of [23].

## 4.2.4 WIRELESS DOWNLINK STREAMING

Different from the uplink case, transmissions in the downlink are orthogonal to each other, thus it is desirable to allow simultaneous transmissions of multiple video users. The resource constraint in the downlink case is the maximum peak transmission power at the base station. The objective here is to determine the transmission power functions, $p_n(t)$, of each user $n$ during time $t \in [0, T]$, such that the total user utility (measured in video quality) is maximized.

Following the framework described in Section 4.2.1, the first step is to perform average resource allocation. For the downlink case, we will allocate the transmission power to each user, subject to the total transmission power constraint (for video users) at the base station, $P_{\max}^{base}$. Since there is no mutual interference, the transmissions of the voice users need not be taken into consideration when determining the achievable rates of the video users.

At this stage, we will temporality assume that each user $n$ will transmit at a fixed power level $p_n$ throughout the time segment $[0, T]$. Hence, user $n$'s total throughput within $[0, T]$ is

$$r_n(p_n) = T W \log_2 \left( 1 + \frac{h_n p_n}{n_0 W} \right). \tag{4.45}$$

The system optimization problem is then

$$\max_{\{p_n \geq 0, \forall n\}} \sum_{n=1}^{N} u_n \left( r_n(p_n) \right), \text{ s.t. } \sum_{n=1}^{N} p_n \leq P_{\max}^{base}. \tag{4.46}$$

Problem (4.46) is a special case of Problem (4.35), and the optimal transmission power per user can be solved using Algorithm 1.

Due to the differences in frame sizes and locations, transmitting at constant power levels is not optimal in terms of meeting the frame delivery deadlines. We can further perform an energy-efficient water-filling power allocation to improve the performance of Problem (4.46). For details, see Section IV of [23].

## 4.3  APPLICATION II: WIRELESS SERVICE PROVIDER COMPETITION

Due to the deregulation of the telecommunication industry, future wireless users are likely to be able to freely choose a provider (or providers) offering the best tradeoff of parameters in real time. This is already happening with some public Wi-Fi connections, where users can connect to wireless access points of their choices, with usage-based payments and no contracts. Despite the common presence of a free public Wi-Fi network, some users may still choose more expensive providers who offer better quality of service.

In this application, we consider a situation where wireless service providers compete to sell limited wireless resources (e.g., frequency bands, time slots, transmission power) to users who are free to choose provider(s). We investigate how providers set prices for the resource, and how users choose the amount of resources they purchase and from which providers. The focus of our study is to characterize the outcome of this interaction. We consider the general case where different users have different utility functions and experience different channel conditions to different service providers.

A proper model for this system is a multi-leader-follower game. The providers announce the wireless resource prices in the first stage, and the users announce their demand for the resource in the second stage. A user's choice is based on providers' prices and its channel conditions. The providers select their prices to maximize their revenues, keeping in mind the impact of their prices on the demand of the users. As in [28], we assume that users pay for the allocated resources instead of the received services. This turns out to be crucial in achieving the globally optimal resource allocation. However, in this section we will first look at the corresponding social welfare optimization problem, as well as a distributed primal-dual algorithm that can achieve the optimal solution of the problem. In Section 6.3, we will revisit this problem and see how to analyze the game theoretical interactions between the competitive providers. A surprising result there is that the equilibrium of the game is actually the same as the optimal solution of the social welfare optimization problem studied here under fairly mild technical assumptions.

### 4.3.1  SYSTEM MODEL

We consider a set $\mathcal{J} = \{1, \ldots, J\}$ of service providers and a set $\mathcal{I} = \{1, \ldots, I\}$ of users. Provider $j \in \mathcal{J}$ has a total of $Q_j$ resource. A user $i \in \mathcal{I}$ can obtain resource from one or more providers, with a demand vector $q_i = (q_{ij}, \forall j \in \mathcal{J})$ and $q_{ij}$ represents the demand from user $i$ to provider $j$. We use $q = (q_i, \forall i \in \mathcal{I})$ to denote the demand vector of all users.

User $i$'s utility function would be $u_i \left( \sum_{j=1}^{J} q_{ij} c_{ij} \right)$, where $c_{ij}$ is the *channel quality offset* for the channel between user $i$ and the base station of provider $j$ (see Example 4.11 and Assumption 4.13), and $u_i$ is an increasing and concave utility function. The communication can be both downlink or uplink, as long as users do not interfere with each other by using orthogonal resources.

Under this model, a user is allowed to purchase from several providers at the same time. For this to be feasible, a user's device might need to have several wireless interfaces. Mathematically, the solution of this model gives an upper bound on best performance of any situation where users are constrained to purchase from one provider alone.

Next we give a concrete example of how our model is mapped into a physical wireless system.

**Example 4.11**  Consider wireless providers operating on orthogonal frequency bands $W_j$, $j \in \mathcal{J}$. Let $q_{ij}$ be the fraction of time that user $i$ is allowed to transmit exclusively on the frequency band of provider $j$, with the constraint $\sum_{i \in \mathcal{I}_j} q_{ij} = 1$, for all $j \in \mathcal{J}$. Furthermore, assume that each user has a peak power constraint $P_i$. We can then define $c_{ij} = W_j \log(1 + \frac{P_i |h_{ij}|^2}{\sigma_{ij}^2 W_j})$, where $h_{ij}$ is the channel gain and $\sigma_{ij}^2$ is the Gaussian noise variance for the channel between user $i$ and network $j$.

Although the $c_{ij}$ channel quality offset factor represents channel capacity in Example 4.11, it can be any increasing function of the channel strength depending on the specific application scenario.

We make the following assumptions throughout this section:

**Assumption 4.12  Utility functions**   For every user $i \in \mathcal{I}$, $u_i(x)$ is differentiable, increasing, and strictly concave in $x$.

**Assumption 4.13  Channel quality offsets and channel gains**   Channel quality offsets $c_{ij}$ are drawn independently from continuous, possibly different utility distributions. In particular, $Prob(c_{ij} = c_{kl}) = 0$ for any $i, k \in \mathcal{I}$ and $j, l \in \mathcal{J}$. The channel quality offset accounts for the effect that buying the same amount of resource from different providers typically has different effects on a user's quality of service. As Example 4.11 shows, channel quality offset $c_{ij}$ may be a function of the channel gain $h_{ij}$ between user $i$ and provider $j$. In this case, the assumption is fulfilled if channel gains are drawn from independent continuous probability distributions (e.g., Rayleigh, Rician, distance-based path-loss model).

**Assumption 4.14  Atomic and price-taking users**   The demand for an atomic user is not infinitely small and can have an impact on providers' prices. Precise characterization of this impact is one of the focuses of our discussions.

### 4.3.2    SOCIAL WELFARE OPTIMIZATION

Next we formulate the social welfare maximization problem, which aims at maximizing the sum of users' utility functions. For clarity of exposition, we define the following notation.

**Definition 4.15    (Effective resource)**    Let $x = (x_i, \forall i \in \mathcal{I})$ be the vector of *effective resources*, where $x_i(q_i) = \sum_{j=1}^{J} q_{ij} c_{ij}$ is a function of user $i$'s demand $q_i = (q_{ij}, \forall j \in \mathcal{J})$.

The social welfare optimization problem (SWO) is:

$$\textbf{SWO} : \text{maximize } u(x) = \sum_{i \in \mathcal{I}} u_i(x_i) \tag{4.47}$$

$$\text{subject to } \sum_{j \in \mathcal{J}} q_{ij} c_{ij} = x_i, \ \forall i \in \mathcal{I}, \tag{4.48}$$

$$\sum_{i \in \mathcal{I}} q_{ij} = Q_j, \ \forall j \in \mathcal{J}, \tag{4.49}$$

$$\text{variables } q_{ij}, x_i \geq 0, \ \forall i \in \mathcal{I}, j \in \mathcal{J}. \tag{4.50}$$

For clarity we expressed the SWO in terms of two different variables: effective resource vector $x$ and demand vector $q$, even though the problem can be expressed entirely in terms of $q$. In particular, a vector $q$ uniquely determines a vector $x$ through equations (4.48), i.e., we can write $x$ as $x(q)$. With some abuse of notation we will write $u(q)$ when we mean $u(x(q))$.

Since $u_i(x_i)$ is strictly concave in $x_i$, then $u(x) = \sum_{i \in \mathcal{I}} u_i(x_i)$ is strictly concave in $x$. The feasible region defined by constraints (4.48)–(4.50) is convex. Hence, $u(x)$ has a unique optimal solution $x^*$ subject to constraints (4.48)–(4.50).

However, notice that even though $u_i(\cdot)$ is strictly concave in $x_i$, it is not strictly concave in the demand vector $q_i$. Hence, SWO is non-strictly concave in $q$. It is well known that a non-strictly concave maximization problem might have several different global optimizers (several different demand vectors $q$ in our case) [29, 30]. In particular, one can choose $c_{ij}$'s, $Q_j$'s, and $u_i(\cdot)$'s in such a way that a demand maximizing vector $q^*$ of SWO is not unique. However, we can show that such cases arise with zero probability whenever channel offsets factors $c_{ij}$'s are independent random variables drawn from continuous distributions (see Assumption 4.13). For details, see Section III of [31].

Given an optimal demand vector $q^*$ of the SWO problem, there exists a unique corresponding Lagrange multiplier vector $p^*$, associated with the resource constraints of $J$ providers [20]. This $p^*$ actually can be interpreted as the prices announced by the providers, which will be useful for understanding the following primal-dual algorithm.

### 4.3.3    PRIMAL-DUAL ALGORITHM

The previous analysis assumes that a centralized decision maker can perform network optimization with complete network information. This may not be true in practice. Next we present a distributed primal-dual algorithm where providers and users only know local information and make local decisions in an iterative fashion. We will show that the primal-dual algorithm converges to a set containing the optimal solution of SWO. We can further show that this set contains only the unique optimal solution in most cases, regardless of the values of the updating rates. We first present the algorithm, and then the proof of its convergence.

We will consider a continuous-time algorithm, where all the variables are functions of time. For compactness of exposition, we will sometimes write $q_{ij}$ and $p_j$ when we mean $q_{ij}(t)$ and $p_j(t)$, respectively. Their time derivatives $\frac{\partial q_{ij}}{\partial t}$ and $\frac{\partial p_j}{\partial t}$ will often be denoted by $\dot{q}_{ij}$ and $\dot{p}_j$. We denote by $q^*$ and $p^*$ the unique maximizer of SWO and the corresponding Lagrange multiplier vector, respectively.

To simplify the notation, we denote by $f_{ij}(t)$ or simply $f_{ij}$ the marginal utility of user $i$ with respect to $q_{ij}$ when his demand vector is $q_i(t)$:

$$f_{ij} = \frac{\partial u_i(q_i)}{\partial q_{ij}} = c_{ij} \frac{\partial u_i(x)}{\partial x}\bigg|_{x=x_i=\sum_{j=1}^{J} q_{ij} c_{ij}}. \tag{4.51}$$

We will use $f_{ij}^*$ to denote the value of $f_{ij}(t)$ evaluated at $q_i^*$, the maximizing demand vector of user $i$. So, $f_{ij}^*$ is a constant that is equal to a user's marginal utility at the global optimal solution of the SWO problem, as opposed to $f_{ij}(t)$ which indicates marginal utility at a particular time $t$. We also define $\nabla u_i(q_i) = (f_{ij}, \forall j \in \mathcal{J})$ and $\nabla u_i(q_i^*) = (f_{ij}^*, \forall j \in \mathcal{J})$, where all the vectors are column vectors.

We define $(x)^+ = \max(0, x)$ and

$$(x)_y^+ = \begin{cases} x & y > 0 \\ (x)^+ & y \leq 0. \end{cases}$$

Motivated by the work in [32], we consider the following standard *primal-dual variable update algorithm*:

$$\dot{q}_{ij} = k_{ij}^q \left( f_{ij} - p_j \right)_{q_{ij}}^+, \quad \forall i \in \mathcal{I}, \forall j \in \mathcal{J}, \tag{4.52}$$

$$\dot{p}_j = k_j^p \left( \sum_{i=1}^{I} q_{ij} - Q_j \right)_{p_j}^+, \quad \forall j \in \mathcal{J}. \tag{4.53}$$

Here $k_{ij}^p$'s and $k_j^p$'s are the constants representing update rates. The update rule ensures that, when a variable of interest ($q_{ij}$ or $p_j$) is already zero, it will not become negative even when the direction of the update (i.e., quantity in the parenthesis) is negative. The tuple $(q(t), p(t))$ controlled by equations (4.52) and (4.53) will be referred to as the *solution trajectory* of the differential equations system defined by (4.52) and (4.53).

The motivation for the proposed algorithm is quite natural. A provider increases its price when the demand is higher than its supply and decreases its price when the demand is lower. A user decreases his demand when a price is higher than his marginal utility and increases it when a price is lower. In essence, the algorithm is following the natural direction of market forces.

One key observation is that these updates can be implemented in a distributed fashion. The users only need to know the prices proposed by the providers. The providers only need to know the demand of the users for their own resource, and not for the resource of other providers. In particular, user $i$ only needs to know his own channel offset parameters $c_{ij}, j \in \mathcal{J}$.

Next we prove the convergence of the algorithm. The first step is construct a lower-bounded La Salle function $V(q(t), p(t))$ as follows:

$$V(q(t), p(t)) = V(t) = \sum_{i,j} \frac{1}{k_{ij}^q} \int_0^{q_{ij}(t)} (\beta - q_{ij}^*) d\beta + \sum_j \frac{1}{k_j^p} \int_0^{p_j(t)} (\beta - p_j^*) d\beta. \quad (4.54)$$

It can be shown that $V(q(t), p(t)) \geq V(q^*, p^*)$, i.e., $V$ is bounded. This ensures that if the function $V$ is non-increasing, it will eventually reach a constant value (which may or may not be the global minimum $V(q^*, p^*)$).

The second step is to show that the value of $V(q(t), p(t))$ is non-increasing for any solution trajectory $(q(t), p(t))$ that satisfies (4.52) and (4.53). More specifically, we can show that the derivative of $V$ *along the solution trajectories of the system,*

$$\dot{V}(t) = \sum_{i,j} \frac{\partial V}{\partial q_{ij}} \dot{q}_{ij} + \sum_j \frac{\partial V}{\partial p_j} \dot{p}_j,$$

is always nonpositive. This will ensure that $(q(t), p(t))$ converge to a set of values that keeps $V(q(t), p(t))$ constant.

**Proposition 4.16**    *The pair* $(q(t), p(t))$ *that satisfies (4.52) and (4.53) converges to the invariant set* $V_L = \{q(t), p(t) : \dot{V}(q(t), p(t)) = 0\}$ *as t goes to* $\infty$.

It is clear that the invariant set $V_L$ contains the solution trajectory that has the value of the unique maximizer of SWO $(q^*(t), p^*(t)) = (q^*, p^*)$ for all $t$, since $\dot{V}(q^*, p^*) = 0$. However, it may contain other points as well. When the trajectory $(q(t), p(t))$ enters the invariant set, it either reaches its minimum (i.e., by converging to the unique equilibrium point $(q^*, p^*)$), or it gets stuck permanently in some limit cycle. In either case, the trajectory will be confined to a subset of $V_L = \{(q(t), p(t)) : \dot{V}(q(t), p(t)) = 0\}$.

The good news is that we can indeed show that the invariant set $V_L$ contains only the equilibrium point $(q^*, p^*)$. This can be done in two steps. First, we show that the set $V_L$ has only one element for the majority of the network scenarios, without any restrictions on the variable update rates. Second, we provide a sufficient condition on the update rates so that the global convergences to the unique equilibrium point are also guaranteed in the remaining scenarios. For details, see [31].

## 4.3.4    NUMERICAL RESULTS

For numerical results, we extend the setup from Example 4.11, where the resource being sold is the fraction of time allocated to the exclusive use of the providers' frequency band, i.e., $Q_j = 1$ for $j \in \mathcal{J}$. We take the bandwidth of the providers to be $W_j = 20$MHz, $j \in \mathcal{J}$. User $i$'s utility function is $a_i \log(1 + \sum_{j=1}^{J} q_{ij} c_{ij})$, where we compute the spectral efficiency $c_{ij}$ from the Shannon formula $\frac{1}{2} W \log(1 + \frac{E_b/N_0}{W} |h_{ij}|^2)$, $q_{ij}$ is the allocated time fraction, $E_b/N_0$ is the ratio of transmit power to thermal noise, and $a_i$ is the individual willingness to pay factor taken to be the same across users. The channel gain amplitudes $|h_{ij}| = \frac{\xi_{ij}}{d_{ij}^{\alpha/2}}$ follow Rayleigh fading, where $\xi_{ij}$ is a Rayleigh distributed random variable with parameter 1, and $\alpha = 3$ is the outdoor power distance loss. We choose the parameters so that the $c_{ij}$ of a user is on average around 3.5Mbps when the distance is 50m, and around 60Mbps when the distance is 5m. The average signal-to-noise ratio $E_b/(N_0 d^\alpha)$ at 5m is around 25dB. We assume perfect modulation and coding choices such that the communication rates come from a continuum of values. The users are uniformly placed in a 200m by 200m area. We want to emphasize that the above parameters are chosen for illustrative purposes only. Our theory applies to any number of providers, any number of users, any type of channel attenuation models, and arbitrary network topologies.

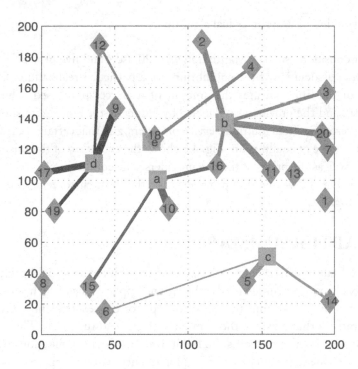

**Figure 4.7:** Example of equilibrium user-provider association. The users are labeled by numbers (1-20), and the providers are labeled by letters ($a$-$e$).

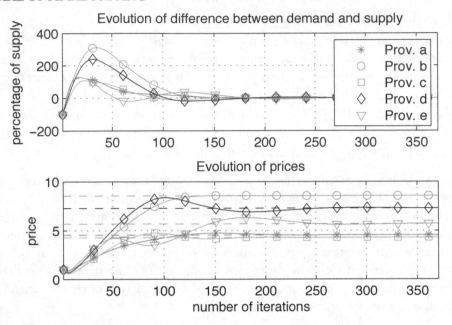

**Figure 4.8:** Evolution of the primal-dual algorithm.

We first consider a single instantiation with 20 users and five providers. In Fig. 4.7, we show the user-provider association at the equilibrium for a particular realization of channel gains, where the thickness of the link indicates the amount of resources purchased. The users are labeled by numbers (1-20), and the providers are labeled by letters (*a-e*). This figure shows two users (12 and 16) requesting resources from multiple service providers, and that certain users (1,7,13, and 8) do not purchase any resource at equilibrium. Fig. 4.8 shows the evolution of the mismatch between supply and demand as well as the prices of the five providers. The equilibrium prices reflect the competition among users: in Fig. 4.7 we see that provider *b* has the most customers, so it is not surprising that its price is the highest, as seen in Fig. 4.8.

## 4.4  CHAPTER SUMMARY

Pricing can facilitate a wireless network operator to achieve the social optimality in a distributed fashion. In this chapter, we discussed the theory and applications of such social optimal pricing.

We started by introducing the basic concepts of convex sets and convex functions, as well as several operations that preserve the convexity of sets and functions. This helps us to define the convex optimization, which concerns the minimization of convex functions over convex sets. Then we moved on with the key theory of this chapter: duality-based distributed algorithms for solving a convex problem. We first introduced the Lagrange dual problem formulation of a primal optimization problem (not necessarily convex), and characterized the KKT necessary conditions under which a primal dual feasible solution pair is optimal for both the primal and dual problems. Most importantly, the dual variables have the nice interpretations of shadow prices, and the dual problem can be solved in a distributed fashion through the subgradient method using shadow prices as coordinating signals.

To illustrate the idea of dual-based distributed optimization, we showed two applications in wireless networks. The first application concerns the resource allocation and scheduling for wireless video streaming in a single cell CDMA network. The problem is rather complicated as the video source optimization is often discrete and thus not convex. However, we decompose the network optimization problem into three phrases: average resource allocation, video source adaptation, and multiuser deadline oriented scheduling. We showed that it is possible to only consider the "semi-elastic" nature of today's video sources to perform average resource allocation via a dual-based resource allocation. We make the discussions more concrete by considering different formulations in both wireless uplink and downlink streaming. In the second application, we consider the resource allocation among multiple wireless service providers serving overlapping areas. As users have the flexibility of purchasing from one or more service providers, the social welfare maximization turns out to be convex but not strictly convex. We design a primal-dual algorithm, which is a generalization of the dual-based algorithm, such that the users and providers can coordinate in a distributed fashion and converge to the unique global optimal solution. For more details especially mathematical proofs related to the two applications, see [23, 31].

## 4.5 EXERCISES

1. Determine the concavity and convexity of the following functions, where $x$ is the variable.

   - *Shannon Capacity of AWGN Channels*

   $$f(x) = \log\left(1 + \frac{x}{\sigma^2}\right), \quad x \in \mathbb{R}^+,$$

   where $\sigma^2$ is the normalized noise power (constant) and $x$ is the received signal power.

   - *Sum-Rate Capacity of OFDM Systems*

   $$f(x_1, ..., x_n) = \sum_{i=1}^{n} \log\left(1 + \frac{x_i}{\sigma_i^2}\right), \quad x_i \in \mathbb{R}^+, \forall i = 1, ..., n,$$

   where $\sigma_i^2$ is the normalized noise power (constant) and $x_i$ is the received signal power on the $i$-th sub-channel.

   - *Bit-Error-Rate (BER) for BPSK Modulation*

   $$f(x) = Q\left(\sqrt{2x}\right), \quad x \in \mathbb{R}^+,$$

   where the $Q$–function is defined as $Q(t) = \frac{1}{\sqrt{2\pi}} \int_t^\infty e^{-t^2/2} dt$. (Note: if a function is neither concave nor convex, then show the region where it is concave or convex.)

2. *Power allocation problem in OFDM systems.* Consider an uplink OFDM system, where a mobile user transmits data to the base station over $n$ sub-channels. Each sub-channel experiences a different channel gain due to the frequency selective fading effect. Let $G_i$ denote the channel gain of sub-channel $i$. The mobile user needs to determine the transmission power on every sub-channel to maximize the sum-rate capacity, subject to several power constraints. Specifically, let $p_i$ denote the transmission power of sub-channel $i$. The power constraints are

$$\text{(i)} \sum_{i=1}^{n} p_i \leq P^{tot}, \quad \text{and} \quad \text{(ii)} \ p_i \leq P^{one}, \quad \forall i,$$

where (i) is the mobile user's total power constraint, and (ii) is the per-channel power constraint. Using the Shannon capacity formula, the sum-rate capacity is

$$f(p_1, ..., p_n) = \sum_{i=1}^{n} \log \left( 1 + \frac{p_i G_i}{\sigma^2} \right).$$

Formulate the above power allocation problem, and study its concavity and convexity. Drive the optimal solution(s).

3. *Network flow problem.* Consider a network of $n$ nodes. Each pair of nodes $(i, j)$ is connected by a directed link $l_{ij}$ from node $i$ to node $j$, with a link capacity $B_{ij}$ (i.e., the maximum allowable flow from node $i$ to $j$) and a serving cost $C_{ij}$ (i.e., the unit cost of the flow along the link $l_{ij}$). Let $I_i \geq 0$ denote the volume of external flows entering the network through node $i$, and $O_i \geq 0$ denote the volume of flows flowing out of the network through node $i$. Without loss of generality, we suppose that the whole network is balanced, i.e., $\sum_{i=1}^{n} I_i = \sum_{i=1}^{n} O_i$. The network flow problem is to schedule the flows among different links to minimize the total serving cost, subject to the link capacity constraint and the node flow balance constraint. Specifically, let $x_{ij}$ denote the flow from node $i$ to $j$. Then, the link capacity constraints are

$$x_{ij} \leq B_{ij}, \quad \forall i, \forall j;$$

The node balance constraints are

$$I_i + \sum_{j=1}^{n} x_{ji} = O_i + \sum_{j=1}^{n} x_{ij}, \quad \forall i;$$

Denote $\boldsymbol{x} = \{x_{ij}, \forall i, \forall j\}$. The total cost across the network is

$$f(\boldsymbol{x}) = \sum_{i,j=1}^{n} C_{ij} \cdot x_{ij}.$$

Give the detailed formulation such a network flow problem (which is a linear programming problem). Derive the optimal solution(s).

CHAPTER 5

# Monopoly and Price Discriminations

In this chapter, we will move away from social optimal pricing and study the issue of profit maximization instead. In particular, we will investigate the case where a single service provider dominates the market. The provider can charge a single optimized price to all the consumers, or he can charge different prices based on the consumer types if such information is available. Such price differentiation may significantly improve the provider's profit.

We first introduce the theory of monopoly pricing and price discriminations, and illustrate the theory through two examples. In the first example, we consider the revenue maximization of a cellular operator, who faces users of different channel conditions and hence different capabilities of utilizing the resources. In the second example, we discuss how a service provider should optimize partial price differentiation, when it is constrained to charge only a limited number of different prices to consumers.

## 5.1 THEORY: MONOPOLY PRICING

In this section, we cover the basic concepts of monopoly pricing. Our discussions follow closely those in [16, 17].

### 5.1.1 WHAT IS MONOPOLY?

Before discussing the basic theory in monopoly pricing, we need to define what "monopoly" means. Etymology suggests that a "*monopoly*" is a *single seller*, the only firm in its industry. But such a vague answer may cause serious confusion. Consider Apple Inc., which is obviously the only firm that sells iPhone; however, Apple is *not* the only firm that sells mobile phones. Thus, whether Apple is a single seller depends on how narrowly we define the market.

In order to avoid such confusions, we will use a different definition relying on the *monopoly power* or *market power*, a widely used concept in economics. As defined in many economic literatures (e.g., [17]), monopoly power or market power is the ability of a firm to affect market prices through its actions. A firm with monopoly power is referred to as a *monopoly* or *monopolist*. More specifically,

**Definition 5.1 Monopoly Power** A firm has monopoly power, if and only if (i) it faces a downward-sloping demand curve for its product, and (ii) it has no supply curve.

The first condition implies that a monopolist is not perfectly competitive. That is, he is able to set the market price so as to shape the demand. The second condition implies that the market price is a consequence of the monopolist's actions, rather than a data to which he must react. By this definition, Apple is obviously a monopoly (in the iPhone market), since it can lower the price of iPhone to increase the sales of iPhones (i.e., the demand curve for iPhone slops downward). The competitive soy farmer who can increase/decrease his output and still sell it all at the going market price is not a monopoly (in the soy market).

In what follows, we will study how a monopolist chooses price and quantity, and what are the profit consequences of these choices.

## 5.1.2   PROFIT MAXIMIZATION BASED ON DEMAND ELASTICITY

Let $P$ denote the market price a monopolist chooses. Let $Q \triangleq D(P)$ denote the downward-sloping demand curve the monopolist faces (or the best quantity the monopolist chooses). A key question is: *how should the monopolist choose a market price to maximize his profit?* We will show that the answer depends greatly on the demand curve the monopolist faces. In particular, it depends on the *price elasticity* of demand defined in Section 3.2.

We first consider the monopolist's total revenue $\pi(P)$ under a particular market price $P$. Formally, we have

$$\pi(P) \triangleq P \cdot Q, \quad \text{where } Q = D(P). \tag{5.1}$$

It is easy to check that $\pi(P)$ is a concave function of $P$. Therefore, the optimal price $P^*$ that maximizes $\pi(P)$ is given by the first-order condition:

$$\frac{d\pi(P)}{dP} = Q + P \cdot \frac{dQ}{dP} = 0, \tag{5.2}$$

which leads to the following optimality condition:

$$\frac{P \cdot \triangle Q}{Q \cdot \triangle P} + 1 = 0, \tag{5.3}$$

where $\triangle Q$ and $\triangle P$ are small changes in quality and price, respectively.

Next we show that the above revenue maximization problem is closely related to the following problem: *how much does the monopolist have to lower his price to sell one more product?* The answer leads to the price elasticity of demand defined in Section 3.2. Recall that the price elasticity of demand is defined as the change in demand that results from one unit increase in price, and given by the formula:

$$\eta \triangleq \frac{\triangle Q/Q}{\triangle P/P} = \frac{P \cdot \triangle Q}{Q \cdot \triangle P}, \tag{5.4}$$

or equivalently,

$$\triangle P = \frac{P \cdot \triangle Q}{Q \cdot \eta}, \tag{5.5}$$

which shows how much the market price must change to sell additional $\triangle Q$ of product. Note that to sell an extra product, the change in price $\triangle P$ must be negative, which can be confirmed by the fact that $\eta$ is always negative. Thus, we can also write the absolute value of $\triangle P$ as $|\triangle P| = \frac{P \cdot \triangle Q}{Q \cdot |\eta|} = -\frac{P \cdot \triangle Q}{Q \cdot \eta}$.

Now we consider the consequences of selling an additional product. That is, how much the monopolist's total revenue changes by selling an additional product. Specifically, there are two factors affecting the monopolist's revenue $\pi$. On one hand, the monopolist gains an additional revenue $P \cdot \triangle Q$ by selling an additional unit $\triangle Q$ of product at price $P$. On the other hand, the monopolist suffers a revenue loss $|\triangle P| \cdot Q$, since the price for the previous $Q$ products is decreased by $|\triangle P|$. Thus, the net change in the monopolist's revenue is

$$\triangle \pi \triangleq P \cdot \triangle Q - |\triangle P| \cdot Q. \tag{5.6}$$

Substitute (5.5) into (5.6), we can rewrite the revenue change as

$$\triangle \pi = P \cdot \triangle Q - \frac{P \cdot \triangle Q}{Q \cdot |\eta|} \cdot Q = P \cdot \triangle Q \cdot \left(1 - \frac{1}{|\eta|}\right), \tag{5.7}$$

which shows how much the monopolist's revenue changes by selling an additional unit $\triangle Q$ of product. Note that when $\triangle Q = 1$, it is essentially the monopolist's *marginal revenue* (MR), i.e., the change in his revenue by selling one extra product.

The formula (5.7) shows that $\triangle \pi < 0$ if $|\eta| < 1$. This implies that a monopolist would never lower the price (or increase the quantity equivalently) when $|\eta| < 1$. In other words, a monopolist must operate on a market price or quantity such that $|\eta| \geq 1$. If in addition there is no other cost, the optimal price or quantity satisfies $\triangle \pi = 0$, which implies that $|\eta| = 1$ or $1 + \eta = 0$. This is obviously equivalent to the first-order condition in (5.2). Note that if there is certain cost (e.g., the producing cost), the optimal price or quantity satisfies $\triangle \pi = \triangle C$, i.e., the change in revenue equals to the change in cost.

When $|\eta| > 1$, we say that the demand curve is *elastic*; when $|\eta| < 1$ we say that the demand curve is *inelastic*. An immediate observation is that a profit-maximization monopolist will increase the price whenever the demand curve is inelastic. Thus, our conclusion for a monopolist's operation is given in the following theorem [17].

**Theorem 5.2**  *A monopolist always operates on the elastic portion of the demand curve.*

## 5.2  THEORY: PRICE DISCRIMINATIONS

The analysis of monopoly pricing in Section 5.1 assumes that the monopolist will sell all of products at a single price. This section deals with a monopolist that can engage in *price discrimination* (or price differentiation), i.e., charging different prices for the same product. The goal of price discrimination is to raise the monopolist's revenue by reducing consumer surplus.

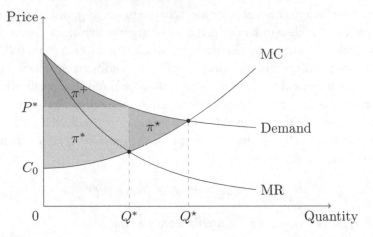

**Figure 5.1:** Increasing the monopolist's profit by eliminating consumer surplus. When charging a single monopoly price $P^*$ to all consumers, the monopolist's profit is shown by the shaded area $\pi^*$ and consumer surplus is shown by the shaded area $\pi^+$. Suppose the monopolist charges each consumer the most that he would be willing to pay for each product that he buys, the monopolist's profit is now $\pi^* + \pi^+$ (if still selling $Q^*$), and the consumers get zero surplus.

Basically, with price discrimination, the monopolist can either charge different prices to a single consumer (for different units of products), or charge uniform but different prices to different groups of consumers. In this section, we will discuss the motivations, conditions, and means of price discrimination.

### 5.2.1  AN ILLUSTRATIVE EXAMPLE

We first show that it is possible for a monopolist to improve his revenue and eliminate consumer surplus by price discrimination.

Consider a simple example with a monopolist facing a downward-sloping demand curve $Q = D(P)$ and a production cost $C(Q)$. The demand curve (D), marginal revenue (MR), and marginal cost (MC) are illustrated in Figure 5.1. The marginal revenue function intersects the marginal cost function when the monopolist chooses the monopoly quantity $Q^*$ and sells each product at the monopoly price $P^*$. The monopolist's total profit is equal to the area labeled $\pi^*$, and the consumer surplus is equal to the area labeled $\pi^+$.

Note that by charging the single monopoly price $P^*$ to all consumers, the monopolist does not collect all the consumer surplus, since consumers are still left with a surplus of $\pi^+$. Now suppose that the monopolist can charge different prices for different units of products. Then, he can collect the consumer surplus $\pi^+$ by charging the demand price $P(D)$ for each successive unit of product, i.e., charging each consumer the most that he would be willing to pay for each additional product that he

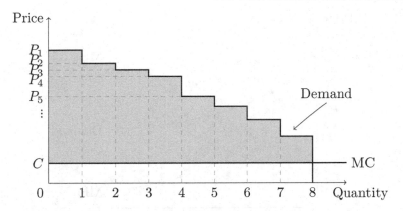

**Figure 5.2:** Under first-degree price discrimination, the consumer is charged a price $P_1$ for the first product he purchases, $P_2$ for the second product he purchases, and so on.

buys. Suppose in addition that the monopolist increases the monopoly quantity $Q^*$ to $Q^\star$, i.e., the interaction of the marginal cost and the demand curve. Then, he cannot only collect the consumer surplus in the area $\pi^+$, but also the surplus in the area $\pi^\star$. In this case, the monopolist collects all the social surplus $\pi^* + \pi^+ + \pi^\star$, which is also the maximum social surplus the monopolist can achieve. This is essentially the first-degree price discrimination, which will be discussed later.

The above example shows that a monopolist can increase its profit by charging different prices to a consumer or to different consumers. Next we will show that the amount of additional profit the monopolist can extract from consumers depends on the information he has about the consumers. As a result, there are three types of price discriminations: first-, second-, and third-degree, which will be discussed in the following sections.

### 5.2.2  FIRST-DEGREE PRICE DISCRIMINATION

With the **first-degree price discrimination**, or *perfect price discrimination*, the monopolist charges each consumer the most that he would be willing to pay for each product that he buys [17]. It requires that the monopolist knows exactly the maximum price that every consumer is willing to pay for each product, i.e., the full knowledge about every consumer demand curve. In this case, the monopolist captures all the market surplus, and the consumer gets zero surplus.

Figure 5.2 illustrates the first-degree price discrimination, where the consumer is willing to pay a maximum price $P_1$ for the first product, $P_2$ for the second product, and so on. When the price is between $[P_2, P_1]$, the total demand is 1; while when the price is between $[P_4, P_3]$, the total demand is 3; and so on. The demand curve can be represented by the downward stepped curve shown in the figure. Under the first-degree price discrimination, the consumer is charged by his maximum willingness to pay for successive products, i.e., $P_1$ for the first product, $P_2$ for the second product, and so on. Obviously, the monopolist captures all the market surplus (shown shaded).

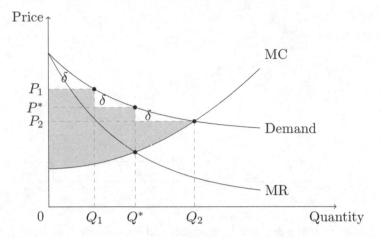

**Figure 5.3:** Under second-degree price discrimination, the consumer is charged a price $P_1$ for the first block (from 0 to $Q_1$) of products, $P^*$ for the second block (from $Q_1$ to $Q^*$), and $P_2$ for the third block (from $Q^*$ to $Q_2$).

In practice, however, it is difficult or even impossible for the monopolist to obtain the complete demand information. Thus, the first-degree price discrimination is primarily theoretical and seldom exists in reality.

### 5.2.3   SECOND-DEGREE PRICE DISCRIMINATION

With the **second-degree price discrimination**, or *declining block pricing*, the monopolist offers a *bundle of prices* to the consumers, with different prices for different blocks of units [17]. Recall that in the first-degree price discrimination, a different price is set for every different unit. In this sense, the second-degree price discrimination can be viewed as a more limited version of the first-degree price discrimination.

Figure 5.3 illustrates a second-degree price discrimination, where the monopolist offers a bundle of prices $\{P_1, P^*, P_2\}$ with $P_1 > P^* > P_2$ to the consumer. The price $P_1$ is for the first block (the first $Q_1$ units) of products, $P^*$ is for the second block (from $Q_1$ to $Q^*$) of products, and $P_2$ is for the third block (from $Q^*$ to $Q_2$). That is, the consumer pays $P_1$ for each unit (of product) up to $Q_1$ units, $P^*$ for each unit between $Q_1$ and $Q^*$ units, and $P_2$ for each unit between $Q^*$ and $Q_2$ units. In effective, we can view the monopolist offers a discount $\frac{P^*}{P_1}$ for the purchasing quantity above $Q_1$, and an additional discount $\frac{P_2}{P^*}$ for the purchasing quantity above $Q^*$. Without price discrimination, the monopolist's maximum profit is $P^*Q^* - C(Q^*)$. With second-degree price discrimination, the monopolist's profit is shown shaded, which is obviously larger than $P^*Q^* - C(Q^*)$. Furthermore, the monopolist's profit with the second-degree price discrimination is less than that with the first-degree price discrimination, and the gap is shown as summation of the blank regions denoted by $\delta$. We can see that if the number of prices the monopolist offers is sufficiently large, and the region of each shaded block is sufficiently close to the demand curve, then the second-degree price discrimination converges effectively to the first-degree price discrimination (i.e., the blank regions vanish).

It is worth noting that the second-degree price discrimination does not require the monopolist to know the complete information of every consumer demand curve. For example, the second-degree price discrimination illustrated in Figure 5.3 requires some particular points on the consumer demand curve only, i.e., $(Q_1, P_1)$, $(Q_2, P_2)$, and $(Q^*, P^*)$. Obviously, the more information the monopolist knows, the higher profit he can absorb from the consumer.

### 5.2.4  THIRD-DEGREE PRICE DISCRIMINATION

A monopolist that performs the first- or second-degree price discrimination knows something about the demand curve of every *individual* consumer, and benefits from this information by charging the consumer different prices. A natural question is whether (and how, if so) the monopolist discriminates the price to increase his profit, if he has no information on the individual demand curve (but knows from experience that different groups of consumers have different total demand curves)?

The answer is YES, and it actually leads to the third and the most common form of price discrimination, the **third-degree price discrimination**, or *multi-market price discrimination* [17]. Simply speaking, third-degree price discrimination usually occurs when a monopolist faces two (or more) *identifiably* different groups of consumers having different (downward-sloping) total demand curves, and knows the total demand curve of every group but not the individual demand curve of every consumer. In this case, the monopolist can potentially increase his profit by setting different prices for different groups.

To apply third-degree price discrimination, the monopolist first uses some characteristic of consumers to segment consumers into groups. Then he picks different prices for the different groups that maximize his profit. In this process, it is implicitly assumed that the monopolist is able to sort consumers into groups (i.e., identify the type of each consumer), and thus consumers in the group with a higher price cannot purchase in the lower-priced market. A simple example of this kind of price discrimination is that the Disney Park offers different ticket prices to children, adults, and elders.

To show how a monopolist discriminates among groups of consumers, we consider a simple scenario, where the monopolist sorts consumers into two groups (two markets). The total demand curves of different markets are different. The monopolist needs to decide the price $P_i$ for each market $i \in \{1, 2\}$ (and therefore the sales $Q_i = D_i(P_i)$ in each market $i$). To maximize his own profit, the monopolist must decide:

- Whether to charge the same price or different prices in different markets?

- Which market should get the lower price if the firm charges different prices?

- What is the relation between the prices of two markets?

Under prices $P_1$ and $P_2$, the monopolist's total profit $\pi(P_1, P_2)$ is given by

$$\pi(P_1, P_2) \triangleq P_1 \cdot Q_1 + P_2 \cdot Q_2 - C(Q_1 + Q_2), \quad \text{where } Q_i = D_i(P_i). \tag{5.8}$$

It is easy to check that $\pi(P_1, P_2)$ is a concave function of vector $(P_1, P_2)$, and therefore the optimal price vector $(P_1^*, P_2^*)$ that maximizes $\pi(P_1, P_2)$ is given by the first-order condition:

$$\frac{\partial \pi(P_1, P_2)}{\partial P_i} = Q_i + P_i \cdot \frac{dQ_i}{dP_i} - C'(Q_1 + Q_2) \cdot \frac{dQ_i}{dP_i} = 0, \quad i = 1, 2, \tag{5.9}$$

where $C'(Q_1 + Q_2) = \frac{dC(Q)}{dQ}|_{Q=Q_1+Q_2}$. It leads to the following optimality condition:

$$C'(Q_1 + Q_2) = P_i + Q_i \cdot \frac{dP_i}{dQ_i} = P_i \cdot \left(1 - \frac{1}{|\eta_i|}\right), \quad i = 1, 2, \tag{5.10}$$

where $\eta_i \triangleq \frac{P_i}{Q_i} \frac{dQ_i}{dP_i}$ is the price elasticity of market $i$. Note that the right-hand side in (5.10) is in fact the marginal revenue in each market $i$. Thus, the above condition suggests that under the optimality, the marginal revenue in each market $i$ equals to the marginal cost. This further leads to

$$P_1 \cdot \left(1 - \frac{1}{|\eta_1|}\right) = P_2 \cdot \left(1 - \frac{1}{|\eta_2|}\right). \tag{5.11}$$

Intuitively, we can see from (5.10) and (5.11) that under the optimality, the monopolist sets an optimal price vector $(P_1^*, P_2^*)$ such that the marginal revenues in all market are the same, and all equal to the marginal cost.

The optimality conditions in (5.10) and (5.11) provider answers to the above three questions. First, we have $P_1 \neq P_2$ as long as $\eta_1 \neq \eta_2$. That is, the monopolist will charge different prices when different groups of customers have different elasticities. Second, we have: (i) $P_1 < P_2$ if $|\eta_1| > |\eta_2|$, and (ii) $P_1 > P_2$ if $|\eta_1| < |\eta_2|$. That is, the market with the higher price elasticity will get a lower price. Third, the relation between the prices of two markets is given by (5.11).

Figure 5.4 provides a graphic interpretation for the above optimal solution. The MR1 and MR2 curves are the marginal revenue curves in both markets, and the D1 and D2 curves are the demand curves in both markets. The MR curve is obtained by summing MR1 and MR2 horizontally. That is, for any price, read the corresponding quantities of MR1 and MR2, and then add these to get the corresponding quantity on MR.

The monopolist, to maximize his profit, can equalize his marginal cost and both marginal revenues by choosing the quantity where his marginal cost curve MC crosses the MR curve (see from (5.10) and (5.11)). This means that he produces a total of $Q_1 + Q_2$ products, so that his marginal cost is 5 per unit of product. He sells $Q_1$ of these products in market 1 and $Q_2$ in market 2, such that his marginal revenue is 5 per unit in each market. Once the monopolist chooses $Q_1$ and $Q_2$, the price for each market is implied given by the inverse demand function, i.e., $P_i = D_i^{-1}(Q_i)$, $i = 1, 2$. From Figure 5.4, we can find that the prices for different markets are different, and the

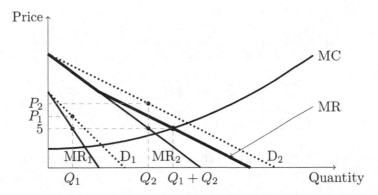

**Figure 5.4:** Under third-degree price discrimination, the monopolist charges a lower price $P_1$ from market 1 (with a higher price elasticity), and a higher price $P_2$ from market 2 (with a lower price elasticity), so that he achieves the same marginal revenues from both markets.

market with the lower price elasticity (market 2 in Figure 5.4) gets the higher price ($P_2$ in Figure 5.4).

We can now summarize the necessary conditions to make the third-degree price discrimination profitable [17].

- **Monopoly power**—The firm must have the monopoly power (or market power) to affect market price, which means that there is no price discrimination in perfectly competitive markets.

- **Market segmentation**—The firm must be able to split the market into different groups of consumers, and also be able to identify the type of each consumer.

- **Elasticity of demand**—There must be a different price elasticity of demand for different markets. This allows the firm to charge a higher price to the market with a relatively inelastic demand and a lower price to those with a relatively elastic demand. The firm will then be able to extract more consumer surplus which will lead to an additional profit.

## 5.3   APPLICATION I: CELLULAR NETWORK PRICING

### 5.3.1   NETWORK MODEL

We consider a cellular operator who owns a single base station and a total of $B$ Hz wireless spectrum. The operator will allocate different parts of the spectrum to different cellular users (such as in the FDMA or OFDMA systems), and charge users accordingly. As shown in Fig. 5.5, we model the interactions between the operator and the users as a two-stage Stackelberg game. In Stage I, the operator determines the price $p$ (per unit bandwidth) to maximize its profit. In Stage II, each user

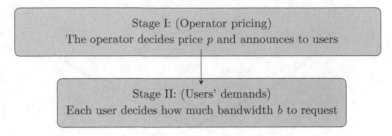

**Figure 5.5:** Two-stage Stackelberg game between the operator and users in Section 5.3.1.

decides how much bandwidth to purchase to maximize its payoff. We next solve this two-stage Stackelberg game by backward induction.

## 5.3.2  USERS' BANDWIDTH DEMANDS IN STAGE II

Different users experience different channel conditions to the base stations due to different locations, and thus achieve different data rates when using the same amount of bandwidth. We consider that a user has fixed transmission power $P$ per unit bandwidth (e.g., power spectrum density constraint) and his average channel gain is $h$. Without interfering with other users, the user's *spectrum efficiency* is thus

$$\theta = \log_2(1 + \text{SNR}) = \log_2\left(1 + \frac{Ph}{n_0}\right),$$

where $n_0$ is the background noise power density. By obtaining $b$ Hz of spectrum, its achieved data rate is $\theta b$ bits per second. As users have different channel gains, they perceive different spectrum efficiency $\theta$. We further normalize $\theta$ to the range [0, 1] (see Fig. 5.6). For example, an indoor user may get a very poor cellular signal reception and hence a small $\theta$. This has been a major issue for most cellular technologies, and is particularly serious with the latest 4G cellular systems operating at higher frequency bands with poor penetrating capabilities through walls. For the simplicity of illustration, we assume that $\theta$ is uniformly distributed. We also normalize the total user population to be 1.

For a user with a spectrum efficiency $\theta$, it obtains a utility $u(\theta, b)$ when achieving data rate $\theta b$, i.e.,

$$u(\theta, b) = \ln(1 + \theta b).$$

Such utility is commonly used in economic literature to denote the diminishing return of getting additional resource. The user needs to pay a linear payment $pb$ to the operator, where the price $p$ is announced by the operator in Stage I. The user's *payoff* is the difference between its utility and payment, i.e.,

$$\pi(\theta, b, p) = \ln(1 + \theta b) - pb. \tag{5.12}$$

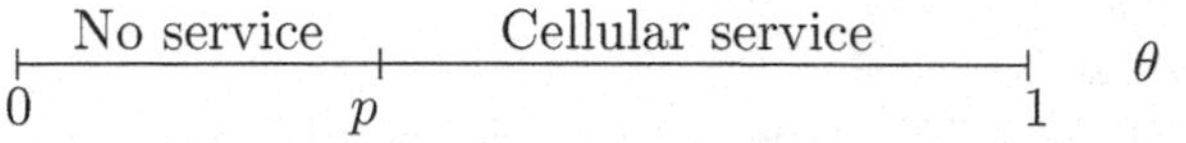

Figure 5.6: Distribution of users' spectrum efficiencies $\theta$.

The optimal value of bandwidth demand that maximizes the user's payoff is

$$b^*(\theta, p) = \begin{cases} \frac{1}{p} - \frac{1}{\theta}, & \text{if } p \leq \theta, \\ 0, & \text{otherwise,} \end{cases} \tag{5.13}$$

which is decreasing in $p$ and increasing in $\theta$ (if $p \leq \theta$). The user's maximum payoff is

$$\pi(\theta, b^*(\theta, p), p) = \begin{cases} \ln\left(\frac{\theta}{p}\right) - 1 + \frac{p}{\theta}, & \text{if } p \leq \theta, \\ 0, & \text{otherwise,} \end{cases} \tag{5.14}$$

which is always nonnegative and is increasing in $\theta$.

### 5.3.3 OPERATOR'S PRICING IN STAGE I

Next we consider the operator's optimal choice of price $p$ in Stage I. To achieve a positive profit, the operator needs to set $p \leq \max_{\theta \in [0,1]} \theta = 1$, so that at least some user purchases some positive bandwidth in Stage II. The fraction of users subscribing to the cellular service is $1 - p$ as shown in Fig. 5.6. The total user demand is

$$Q(p) = \int_p^1 \left(\frac{1}{p} - \frac{1}{\theta}\right) d\theta - \frac{1}{p} - 1 + \ln p, \tag{5.15}$$

which is a decreasing function of $p$. On the other hand, the operator has a limited bandwidth supply $B$, and thus can only satisfy a demand no larger than $B$.

The operator chooses price $p$ to maximize its revenue, i.e.,

$$\max_{0 < p \leq 1} \min\left(Bp, pQ(p)\right). \tag{5.16}$$

Notice that the first term in the min operation of (5.16) is increasing in $p$, while the second term is decreasing in $p$ since

$$\frac{dpQ(p)}{dp} = \ln p < 1.$$

By also checking the two terms' values at the boundary values of $p$, we can conclude that the optimal solution to Problem (5.16) is unique and the two terms are equal at the optimality.

**Theorem 5.3** *The equilibrium price $p^*$ is the unique solution of the following equation:*

$$B = \frac{1}{p^*} - 1 + \ln p^*. \tag{5.17}$$

Notice that all users with a spectrum efficiency $\theta$ less than $p^*$ will not receive the cellular service. When the total bandwidth $B$ is small, the equilibrium price $p^*$ is close to 1 and thus most users will not get service. This motivates the operator to adopt the so-called femtocell service to improve the users' channel conditions and hence the operator's revenue [33].

## 5.4   APPLICATION II: PARTIAL PRICE DIFFERENTIATION

### 5.4.1   SYSTEM MODEL

We consider a network with a total of $S$ units of (infinitely) divisible resource (which can be in the form of rate, bandwidth, power, time slot, etc.). The resource is allocated by a monopolistic service provider to a set $\mathcal{I} = \{1, \ldots, I\}$ of user groups. Each group $i \in \mathcal{I}$ has $N_i$ homogeneous users[1] with the same utility function:

$$u_i(s_i) = \theta_i \ln(1 + s_i), \tag{5.18}$$

where $s_i$ is the allocated resource to one user in group $i$ and $\theta_i$ represents the willingness-to-pay of group $i$. The logarithmic utility function is commonly used to model the proportionally fair resource allocation in communication networks. Without loss of generality, we assume that $\theta_1 > \theta_2 > \cdots > \theta_I$. In other words, group 1 contains users with the highest valuation, and group $I$ contains users with the lowest valuation.

We will assume that the service provider knows each user's utility function. The significance of studying the complete information is two-fold. It serves as the benchmark of practical designs and provides important insights for the incomplete information analysis as in [34].

The interactions between the service provider and the users can be characterized as a two-stage model. The service provider announces the pricing scheme in Stage 1, and users respond with their demands in Stage 2. The users want to maximize their surpluses by optimizing their demands according to the pricing scheme. The service provider wants to maximize its revenue by setting the right pricing scheme to induce desirable demands from users. Since the service provider has a limited total resource, he must guarantee that the total demand from users is no larger than what he can supply.

Next we will discuss the complete price differentiation, where the service provider can charge each user a different price. We will then discuss the other extreme case, where the service provider can only charge a single price to all users. Finally, we will discuss the case where the service provider can charge multiple but a limited number of prices.

### 5.4.2   COMPLETE PRICE DIFFERENTIATION

We first consider the case of complete information. Since the service provider knows the utility and the identity of each user, it is possible to maximize the revenue by charging a different price to each

---

[1]A special case is $N_i = 1$ for each group, i.e., all users in the network are different.

group of users. The analysis will be based on backward induction, starting from Stage 2 and then moving to Stage 1.

**User's Surplus Maximization Problem in Stage 2**

If a user in group $i$ has been admitted into the network and offered a linear price $p_i$ in Stage 1, then it solves the following surplus maximization problem,

$$\underset{s_i \geq 0}{\text{maximize}} \ (u_i(s_i) - p_i s_i), \tag{5.19}$$

which leads to the following unique optimal demand

$$s_i(p_i) = \left(\frac{\theta_i}{p_i} - 1\right)^+, \quad \text{where } (\cdot)^+ \triangleq \max(\cdot, 0). \tag{5.20}$$

**Service Provider's Pricing and Admission Control Problem in Stage 1**

In Stage 1, the service provider maximizes its revenue by choosing the price $p_i$ and the admitted user number $n_i$ for each group $i$ subject to the limited total resource $S$. The key idea is to perform a Complete Price differentiation ($CP$) scheme, i.e., charging each group with a different price.

$$CP: \quad \underset{p \geq 0, s \geq 0, n}{\text{maximize}} \ \sum_{i \in \mathcal{I}} n_i p_i s_i \tag{5.21}$$

$$\text{subject to} \quad s_i = \left(\frac{\theta_i}{p_i} - 1\right)^+, \quad \forall i \in \mathcal{I}, \tag{5.22}$$

$$n_i \in \{0, \dots, N_i\}, \quad \forall i \in \mathcal{I}, \tag{5.23}$$

$$\sum_{i \in \mathcal{I}} n_i s_i \leq S. \tag{5.24}$$

where $p = (p_i, \forall i \in \mathcal{I})$, $s = (s_i, \forall i \in \mathcal{I})$, and $n = (n_i, \forall i \in \mathcal{I})$. Constraint (5.22) is the solution of the Stage 2 user surplus maximization problem in (5.20). Constraint (5.23) denotes the admission control decision, and constraint (5.24) represents the total limited resource in the network.

Problem $CP$ is not straightforward to solve, since it is a non-convex optimization problem with a non-convex objective function (summation of products of $n_i$ and $p_i$), a coupled constraint (5.24), and integer variables $n$. However, it is possible to convert it into an equivalent convex formulation through a series of transformations, and thus the problem can be solved efficiently.

First, we can remove the $(\cdot)^+$ sign in constraint (5.22) by realizing the fact that there is no need to set $p_i$ higher than $\theta_i$ for users in group $i$; users in group $i$ already demand zero resource and generate zero revenue when $p_i = \theta_i$. This means that we can rewrite constraint (5.22) as

$$p_i = \frac{\theta_i}{s_i + 1} \text{ and } s_i \geq 0, i \in \mathcal{I}. \tag{5.25}$$

Plugging (5.25) into (5.21), then the objective function becomes $\sum_{i \in \mathcal{I}} n_i \frac{\theta_i s_i}{s_i + 1}$. We can further decompose Problem $CP$ in the following two subproblems:

1. *Resource allocation*: for a fixed admission control decision $n$, solve for the optimal resource allocation $s$.

$$CP_1: \quad \underset{s \geq 0}{\text{maximize}} \quad \sum_{i \in \mathcal{I}} n_i \frac{\theta_i s_i}{s_i + 1}$$

$$\text{subject to} \quad \sum_{i \in \mathcal{I}} n_i s_i \leq S. \tag{5.26}$$

Denote the solution of $CP_1$ as $s^* = (s_i^*(n), \forall i \in \mathcal{I})$. We further maximize the revenue of the integer admission control variables $n$.

2. *Admission control*:

$$CP_2: \quad \underset{n}{\text{maximize}} \quad \sum_{i \in \mathcal{I}} n_i \frac{\theta_i s_i^*(n)}{s_i^*(n) + 1} \tag{5.27}$$

$$\text{subject to} \quad n_i \in \{0, \ldots, N_i\} , \quad i \in \mathcal{I}$$

Let us first solve Problem $CP_1$ in $s$. Note that it is a convex optimization problem. By using Lagrange multiplier technique, we can get the first-order necessary and sufficient condition:

$$s_i^*(\lambda) = \left( \sqrt{\frac{\theta_i}{\lambda}} - 1 \right)^+ , \tag{5.28}$$

where $\lambda$ is the Lagrange multiplier corresponding to the resource constraint (5.26).

Meanwhile, we note the resource constraint (5.26) must hold with equality, since the objective is strictly increasing function in $s_i$. Thus, by plugging (5.28) into (5.26), we have

$$\sum_{i \in \mathcal{I}} n_i \left( \sqrt{\frac{\theta_i}{\lambda}} - 1 \right)^+ = S. \tag{5.29}$$

This weighted water-filling problem (where $\frac{1}{\sqrt{\lambda}}$ can be viewed as the water level) in general has no closed-form solution for $\lambda$. However, we can efficiently determine the optimal solution $\lambda^*$ by exploiting the special structure of our problem. Note that since $\theta_1 > \cdots > \theta_I$, then $\lambda^*$ must satisfy the following condition:

$$\sum_{i=1}^{K^{cp}} n_i \left( \sqrt{\frac{\theta_i}{\lambda^*}} - 1 \right) = S, \tag{5.30}$$

for a group index threshold value $K^{cp}$ satisfying

$$\frac{\theta_{K^{cp}}}{\lambda^*} > 1 \text{ and } \frac{\theta_{K^{cp}+1}}{\lambda^*} \leq 1. \tag{5.31}$$

In other words, only groups with indices no larger than $K_{cp}$ will be allocated the positive resource. This property leads to the simple Algorithm 2 to compute $\lambda^*$ and the group index threshold $K^{cp}$: we start by assuming $K^{cp} = I$ and compute $\lambda$. If (5.31) is not satisfied, we decrease $K^{cp}$ by one and recompute $\lambda$ until (5.31) is satisfied. Since $\theta_1 > \lambda(1) = (\frac{n_1}{s+n_1})^2\theta_1$, Algorithm 2 always converges and returns the unique values of $K^{cp}$ and $\lambda^*$. The total complexity is $\mathcal{O}(I)$, i.e., linear in the number of user groups (not the number of users).

---

**Algorithm 2** Solving the Resource Allocation Problem $CP_1$

---

1: **function** $CP(\{n_i, \theta_i\}_{i\in\mathcal{I}}, S)$

2: $\quad k \leftarrow I, \lambda(k) \leftarrow \left(\dfrac{\sum_{i=1}^{k} n_i \sqrt{\theta_i}}{S+\sum_{i=1}^{k} n_i}\right)^2$

3: $\quad$ **while** $\theta_k \leq \lambda(k)$ **do**

4: $\quad\quad k \leftarrow k - 1, \lambda(k) \leftarrow \left(\dfrac{\sum_{i=1}^{k} n_i \sqrt{\theta_i}}{S+\sum_{i=1}^{k} n_i}\right)^2$

5: $\quad$ **end while**

6: $\quad K^{cp} \leftarrow k, \lambda^* \leftarrow \lambda(k)$

7: $\quad$ **return** $K^{cp}, \lambda^*$

8: **end function**

---

With $K^{cp}$ and $\lambda^*$, the solution of the resource allocation problem can be written as

$$s_i^* = \begin{cases} \sqrt{\dfrac{\theta_i}{\lambda^*}} - 1, & i = 1, \ldots, K^{cp}; \\ 0, & \text{otherwise.} \end{cases} \tag{5.32}$$

For the ease of discussions, we introduce a new notion of the *effective market*, which denotes all the groups allocated non-zero resource. For resource allocation problem $CP_1$, the threshold $K^{cp}$ describes the size of the effective market. All groups with indices no larger than $K^{cp}$ are *effective groups*, and users in these groups as *effective users*. An example of the effective market is illustrated in Fig. 5.7.

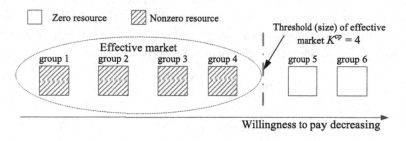

**Figure 5.7:** A 6-group example for effective market: the willingness-to-pay decreases from group 1 to group 6. The effective market threshold can be obtained by Algorithm 2, and is 4 in this example.

Now let us solve the admission control problem $CP_2$. In [34], we show that the objective $R_{cp}(n)$ is strictly increasing in $n_i$ for all $i = 1, \ldots, K^{cp}$, thus it is optimal to admit all users in the effective market. The admission decisions for the groups not in the effective market are irrelevant to the optimization, since those users consume zero resource. Therefore, one of the optimal solutions of Problem $CP_1$ is $n_i^* = N_i$ for all $i \in \mathcal{I}$. Solving Problems $CP_1$ and $CP_2$ leads to the optimal solution of Problem $CP$:

**Theorem 5.4**   *There exists an optimal solution of Problem $CP$ that satisfies the following conditions:*

- *All users are admitted: $n_i^* = N_i$ for all $i \in \mathcal{I}$.*

- *There exist a value $\lambda^*$ and a group index threshold $K^{cp} \leq I$, such that only the top $K^{cp}$ groups of users receive positive resource allocations,*

$$s_i^* = \begin{cases} \sqrt{\dfrac{\theta_i}{\lambda^*}} - 1, & i = 1, \ldots, K^{cp}; \\ 0, & \text{otherwise.} \end{cases}$$

*with the prices*

$$p_i^* = \begin{cases} \sqrt{\theta_i \lambda^*}, & i = 1, \ldots, K^{cp}; \\ \theta_i, & \text{otherwise.} \end{cases}$$

*The values of $\lambda^*$ and $K^{cp}$ can be computed as in Algorithm 2 by setting $n_i = N_i$, for all $i \in \mathcal{I}$.*

Theorem 5.4 provides the right economic intuition: the service provider maximizes its revenue by charging a higher price to users with a higher willingness to pay. It is easy to check that $p_i > p_j$ for any $i < j$. The small willingness-to-pay users are excluded from the markets.

## 5.4.3   SINGLE PRICING SCHEME

We just showed that the $CP$ scheme is the optimal pricing scheme to maximize the revenue under complete information. However, such a complicated pricing scheme is of high implementational complexity. Here we study the other extreme of single pricing scheme. It is clear that the scheme will in general suffer a revenue loss comparing with the $CP$ scheme. We will try to characterize the impact of various system parameters on such revenue loss numerically in Section 5.4.5.

Let us first formulate the Single Pricing ($SP$) problem.

$$\begin{aligned} SP: \quad & \underset{p \geq 0, \; n}{\text{maximize}} \quad p \sum_{i \in \mathcal{I}} n_i s_i \\ & \text{subject to} \quad s_i = \left( \frac{\theta_i}{p} - 1 \right)^+ , \quad \forall i \in \mathcal{I}, \\ & \qquad\qquad\;\; n_i \in \{0, \ldots, N_i\} , \quad \forall \in \mathcal{I}, \\ & \qquad\qquad\;\; \sum_{i \in \mathcal{I}} n_i s_i \leq S. \end{aligned}$$

Comparing with Problem $CP$ in Section 5.4.2, here the service provider charges a single price $p$ to all groups of users. After a similar transformation as in Section 5.4.2, we can show that the optimal single price satisfies the following weighted water-filling condition

$$\sum_{i \in \mathcal{I}} N_i \left( \frac{\theta_i}{p} - 1 \right)^+ = S.$$

Thus, we can obtain the following solution that shares a similar structure as complete price differentiation.

**Theorem 5.5**    *There exists an optimal solution of Problem $SP$ that satisfies the following conditions:*

- *All users are admitted: $n_i^* = N_i$, for all $i \in \mathcal{I}$.*

- *There exist a price $p^*$ and a group index threshold $K^{sp} \leq I$, such that only the top $K^{sp}$ groups of users receive positive resource allocations,*

$$s_i^* = \begin{cases} \frac{\theta_i}{p^*} - 1, & i = 1, 2, \ldots, K^{sp}, \\ 0, & \text{otherwise,} \end{cases}$$

*with the price*

$$p^* = p(K^{sp}) = \frac{\sum_{i=1}^{K^{sp}} N_i \theta_i}{S + \sum_{i=1}^{K^{sp}} N_i}.$$

*The value of $K^{sp}$ and $p^*$ can be computed as in Algorithm 3.*

---

**Algorithm 3** Search the threshold of Problem $SP$

---

1: **function** $SP(\{N_i, \theta_i\}_{i \in \mathcal{I}}, S)$

2:    $k \leftarrow I$, $p(k) \leftarrow \frac{\sum_{i=1}^{k} N_i \theta_i}{S + \sum_{i=1}^{k} N_i}$

3:    **while** $\theta_k \leq p(k)$ **do**

4:        $k \leftarrow k - 1$, $p(k) \leftarrow \frac{\sum_{i=1}^{k} N_i \theta_i}{S + \sum_{i=1}^{k} N_i}$

5:    **end while**

6:    $K^{sp} \leftarrow k$, $p^* \leftarrow p(k)$

7:    **return** $K^{sp}, p^*$

8: **end function**

---

## 5.4.4   PARTIAL PRICE DIFFERENTIATION

For a service provider facing thousands of user types, it is often impractical to design a price choice for each user type. The reasons behind this, as discussed in [35], are mainly high system overheads and customers' aversion. However, the single pricing scheme may suffer a considerable revenue loss compared with the complete price differentiation. How to achieve a good tradeoff between the implementational complexity and the total revenue? In reality, we usually see that the service provider offers only a few pricing plans for the entire users population; we term it as the *partial price differentiation* scheme. In this section, we will answer the following question: if the service provider is constrained to maintain a limited number of prices, $p^1, \ldots, p^J$, $J \leq I$, then what is the optimal pricing strategy and the maximum revenue? Concretely, the Partial Price differentiation ($PP$) problem is formulated as follows.

$$PP: \quad \underset{n_i, p_i, s_i, p^j, a_i^j}{\text{maximize}} \quad \sum_{i \in \mathcal{I}} n_i p_i s_i$$

$$\text{subject to} \quad s_i = \left( \frac{\theta_i}{p_i} - 1 \right)^+, \ \forall i \in \mathcal{I}, \tag{5.33}$$

$$n_i \in \{0, \ldots, N_i\}, \ \forall i \in \mathcal{I}, \tag{5.34}$$

$$\sum_{i \in \mathcal{I}} n_i s_i \leq S, \tag{5.35}$$

$$p_i = \sum_{j \in \mathcal{J}} a_i^j p^j, \tag{5.36}$$

$$\sum_{j \in \mathcal{J}} a_i^j = 1, \ a_i^j \in \{0, 1\}, \forall i \in \mathcal{I}. \tag{5.37}$$

Here $\mathcal{J}$ denotes the set $\{1, 2, \ldots, J\}$. Since we consider the complete information scenario in this section, the service provider can choose the price charged to each group, thus constraints (5.33)–(5.35) are the same as in Problem $CP$. Constraints (5.36) and (5.37) mean that $p_i$ charged to each group $i$ is one of $J$ choices from the set $\{p^j, j \in \mathcal{J}\}$. For convenience, we define *cluster* $\mathcal{C}^j \triangleq \{i \mid a_i^j = 1\}$, $j \in \mathcal{J}$, which is a set of groups charged with the same price $p^j$. We use superscript $j$ to denote clusters, and subscript $i$ to denote groups through this section. We term the binary variables $\boldsymbol{a} \triangleq \{a_i^j, \ j \in \mathcal{J}, \ i \in \mathcal{I}\}$ as the *partition*, which determines which cluster each group belongs to.

Problem $PP$ is a combinatorial optimization problem, and is more difficult than Problem $CP$ and Problem $SP$. On the other hand, we notice that Problem $PP$ includes Problem $CP$ ($J = I$) and Problem $SP$ ($J = 1$) as special cases. The insights we obtained from solving these two special cases in Sections 5.4.2 and 5.4.3 will be helpful to solve the general Problem $PP$.

To solve Problem $PP$, we decompose and tackle it in three levels. In the lowest level-3, we determine the pricing and resource allocation for each cluster, given a fixed partition and fixed resource allocation among clusters. In level-2, we compute the optimal resource allocation among clusters, given a fixed partition. In level-1, we optimize the partition among groups.

## Level-3: Pricing and resource allocation in each cluster

For a fixed partition $a$ and a cluster resource allocation $s \triangleq \{s^j\}_{j \in \mathcal{J}}$, we focus the pricing and resource allocation problems within each cluster $\mathcal{C}^j$, $j \in \mathcal{J}$:

$$\text{Level-3:} \quad \underset{n_i, s_i, p^j}{\text{maximize}} \quad \sum_{i \in \mathcal{C}^j} n_i p^j s_i$$

$$\text{subject to} \quad s_i = \left( \frac{\theta_i}{p^j} - 1 \right)^+, \quad \forall i \in \mathcal{C}^j,$$

$$n_i \leq N_i, \quad \forall i \in \mathcal{C}^j,$$

$$\sum_{i \in \mathcal{C}^j} n_i s_i \leq s^j.$$

Level-3 Subproblem coincides with the $SP$ scheme discussed in Section 5.4.3, since all groups within the same cluster $\mathcal{C}^j$ are charged with a single price $p^j$. We can then directly apply the results in Theorem 3 to solve the Level-3 problem. We denote the effective market threshold for cluster $\mathcal{C}^j$ as $K^j$, which can be computed in Algorithm 3. An illustrative example is shown in Fig. 5.8, where the cluster contains four groups (group 4, 5, 6, and 7), and the effective market contains groups 4 and 5, thus $K^j = 5$. The service provider obtains the following maximum revenue obtained from cluster $\mathcal{C}^j$:

$$R^j(s^j, a) = \frac{s^j \sum_{i \in \mathcal{C}^j, i \leq K^j} N_i \theta_i}{s^j + \sum_{i \in \mathcal{C}^j, i \leq K^j} N_i}. \tag{5.38}$$

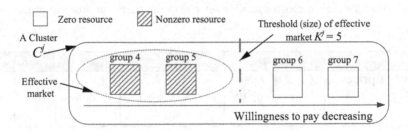

**Figure 5.8:** An illustrative example: the cluster contains four groups, group 4, 5, 6, and 7; and the effective market contains group 4 and 5, thus $K^j = 5$.

## Level-2: Resource allocation among clusters

For a fixed partition $a$, we then consider the resource allocation among clusters.

$$\text{Level-2:} \quad \underset{s^j \geq 0}{\text{maximize}} \quad \sum_{j \in \mathcal{J}} R^j(s^j, a)$$

$$\text{subject to} \quad \sum_{j \in \mathcal{J}} s^j \leq S.$$

In [34], we show that subproblems in Level-2 and Level-3 can be transformed into a complete price differentiation problem under proper technique conditions. Let us denote the optimal value as $R_{pp}(a)$.

**Level-1: cluster partition**

Finally, we solve the cluster partition problem.

$$\text{Level-1: } \quad \underset{a_i^j \in \{0,1\}}{\text{maximize}} \quad R_{pp}(a)$$
$$\text{subject to} \quad \sum_{j \in \mathcal{J}} a_i^j = 1, \; i \in \mathcal{I}.$$

This partition problem is a combinatorial optimization problem. The size of its feasible set is $S(I, J) = \frac{1}{J!} \sum_{t=1}^{J} (-1)^{J+t} C(J, t) t^I$, *Stirling number of the second kind* [36, Chap.13], where $C(J, t)$ is the binomial coefficient. Some numerical examples are given in the third row in Table 5.1. If the number of prices $J$ is given, the feasible set size is exponential in the total number of groups $I$. For our problem, however, it is possible to reduce the size of the feasible set by exploiting the special problem structure. More specifically, the group indices in each cluster should be consecutive at the optimum. This means that the size of the feasible set is $C(I - 1, J - 1)$ as shown in the last row in Table 5.1, and thus is much smaller than $S(I, J)$.

**Table 5.1:** Numerical examples for feasible set size of the partition problem in Level-1

| Number of groups | $I = 10$ | | $I = 100$ | | $I = 1000$ |
|---|---|---|---|---|---|
| Number of prices | $J = 2$ | $J = 3$ | $J = 2$ | $J = 3$ | $J = 2$ |
| $S(I, J)$ | 511 | 9330 | $6.33825 \times 10^{29}$ | $8.58963 \times 10^{46}$ | $5.35754 \times 10^{300}$ |
| $C(I - 1, J - 1)$ | 9 | 36 | 99 | 4851 | 999 |

In [34], we propose an algorithm that solves the three level subproblems in polynomial time and obtains the optimal solution of Problem $PP$. In particular, we show that an optimal partition of Problem $PP$ involves consecutive group indices within clusters, and hence the complexity of solving the level-1 problem can be greatly reduced (comparing to the exhaustive search).

## 5.4.5   NUMERICAL RESULTS

We provide a numerical example to quantitatively study two key questions regarding the performance comparison of different algorithms:

- When is price differentiation most beneficial?

• What is the best tradeoff of partial price differentiation?

**Definition 5.6    (Revenue gain)** We define the revenue gain $G$ of one pricing scheme as the ratio of the revenue difference (between this pricing scheme and the single pricing scheme) normalized by the revenue of single pricing scheme.

We consider a three-group example and three different sets of parameters as shown in Table 5.2. To limit the dimension of the problem, we set the parameters such that the total number of users and the average willingness-to-pay (i.e., $\bar{\theta} = \sum_{i=1}^{3} N_i \theta_i / (\sum_{i=1}^{3} N_i)$) of all users are the same across three different parameter settings. This ensures that the $SP$ scheme achieves the same revenue in three different cases when resource is abundant.

**Table 5.2:** Parameter settings of three-group examples

|        | $\theta_1$ | $N_1$ | $\theta_2$ | $N_2$ | $\theta_3$ | $N_3$ | $\bar{\theta}$ |
|--------|------|------|------|------|------|------|------|
| Case 1 | 9    | 10   | 3    | 10   | 1    | 80   | 2    |
| Case 2 | 3    | 33   | 2    | 33   | 1    | 34   | 2    |
| Case 3 | 2.2  | 80   | 1.5  | 10   | 1    | 10   | 2    |

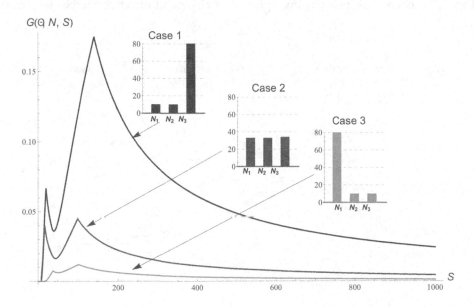

**Figure 5.9:** How the revenue gain $G$ changes with the system parameters in a three-group market.

Figure 5.9 illustrates how the differentiation gain changing changes in resource $S$. We observe that the revenue gain is large only when the high willingness-to-pay users are minorities (e.g., case 1) in the effective market and the resource is limited but not too small ($100 \leq S \leq 150$ in all three cases). When resource $S$ is large enough (e.g., $\geq 150$), the gain will gradually diminish to zero as the resource increases. For each curve in Fig. 5.9, there are two peak points. Each peak point represents a change of the effective market threshold in the benchmark $SP$ scheme, i.e., when the resource allocation to a group becomes zero.

Now let us consider a five-group example with parameters shown in Table 5.3 to illustrate the tradeoff of partial price differentiation. Note that high willingness-to-pay users are minorities here. Figure 5.10 shows the revenue gain $G$ as a function of total resource $S$ under different $PP$ schemes (including the $CP$ scheme as a special case).

**Table 5.3:** Parameter setting of a five-group example

| group index $i$ | 1 | 2 | 3 | 4 | 5 |
|---|---|---|---|---|---|
| $\theta_i$ | 16 | 8 | 4 | 2 | 1 |
| $N_i$ | 2 | 3 | 5 | 10 | 80 |

We enlarge Fig. 5.10 within the range of $S \in [0, 50]$, which is the most complex and interesting part due to several peak points. Similar to Fig. 5.9, we observe $I - 1 = 4$ peak points for each curve in Fig. 5.10. Each peak point again represents a change of effective market threshold of the SP scheme.

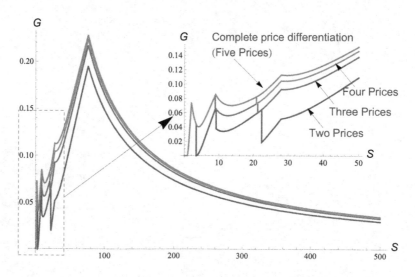

**Figure 5.10:** Revenue gain of a five-group example under different price differentiation schemes.

As the resource $S$ increases from 0, all curves in Fig. 5.10 initially overlap with each other, then the two-price scheme (blue curve) separates from the others at $S = 3.41$, after that the three-price scheme (purple curve) separates at $S = 8.89$, and finally the four-price scheme (dark yellow curve) separates at $S = 20.84$. These phenomena are due to the threshold structure of the $PP$ scheme. When the resource is very limited, the effective markets under all pricing schemes include only one group with the highest willingness to pay, and all pricing schemes coincide with the $SP$ scheme. As the resource increases, the effective market enlarges from two groups to finally five groups.

Figure 5.10 provides the service provider a global picture of how to choose the most proper pricing scheme to achieve the desirable financial target under a certain parameter setting. For example, if the total resource $S = 100$, the two-price scheme seems to be a sweet spot, as it achieves a differential gain of 14.8% comparing to the $SP$ scheme and is only 2.4% worse than the $CP$ scheme with five prices.

## 5.5   CHAPTER SUMMARY

This chapter discussed how a monopoly maximizes its profit through proper pricing mechanisms.

We started by discussing how a monopoly should choose a single profit maximizing price based on the demand elasticity. The key result is that a monopoly will always operate on the elastic portion of the demand curve. Then we look at the monopolist's options when it can charge different prices to the same customer or different customer groups. This leads to three types of price discrimination schemes. In the first-degree price discrimination, the monopoly knows the complete demand information of all customers and performs perfect price discrimination. This unrealistic case provides a theoretical benchmark for other schemes. In the second-degree price discrimination, the monopolist offers a bundle of prices corresponding to different demand quantities, and let the customers choose their best choices. This is often applied when the monopolist knows only limited information of the consumers' demands. In the last third-degree price discrimination, which is one of the most commonly used ones, the monopolist segments the market into several groups, and charges different prices for different groups. This applies for the case that the monopoly knows the total demand for each group, but not the individual demand information. Both second- and third-degree price discrimination induce profit loss comparing to the first degree, and this is inevitable due to the lack of complete information. For more details about the theory, please see [17, 37].

We then introduce two wireless networking examples to illustrate the ideas. We first consider a revenue maximization problem for a cellular operator, who has limited resources and faces users with different channel conditions. The operator wants to sell resources to users who can most effectively utilize the resources and hence have the highest willingness to pay. We can show that under the revenue-maximizing price, some users with poor channel conditions will be reluctant to subscribe to the service. This motivates the cellular provider to deploy the so-called femtocell services to improve the users' channel conditions and hence improve the total revenue [33]. In the second example, we study the revenue-maximizing problem for a monopoly service provider with different number of pricing choices. Our focus is to investigate the tradeoff between the total revenue and the implementational complexity. Among the three pricing differentiation schemes we proposed (i.e., complete, single, and partial), the partial price differentiation is the most general one and includes the other two as special cases. By exploiting the unique problem structure, we are able to design an algorithm that computes the optimal partial pricing scheme in polynomial time, and numerically quantifies the tradeoff between implementational complexity and total revenue. For more details about these two examples, please see [33, 34].

## 5.6 EXERCISES

1. *Power Pricing Problem.* Consider a cognitive radio network where a licensed primary user (PU) shares its spectrum with an unlicensed secondary users (SU). The PU charges the SU a monetary payment based on the received interference from the SU. Let $\pi$ denote the price announced by the PU. Then, the SU decides its transmission power $p \in [0, P_{max}]$ to maximize its utility, which equals the difference between its benefit of data transmission and its payment to the PU, i.e.,

$$U_{su}(p, \pi) = R \cdot \log\left(1 + \frac{p \cdot G_0}{\sigma^2}\right) - p \cdot G_1 \cdot \pi,$$

where $R$ is the unit benefit (revenue) for one unit of achieved data rate, $G_0$ is the channel gain between the SU's transmitter and receiver, $G_1$ is the interference channel gain from the SU's transmitter and the PU's receiver, and $\sigma^2$ is the noise power on the SU's own communication channel. The first term is the total benefit and the second term is the total payment. Show that the SU's power demand curve is

$$Q(\pi) = p^*(\pi) = \frac{R}{G_1 \cdot \pi} - \frac{\sigma^2}{G_0}.$$

The PU's utility is defined as the total payment collecting from the SU, i.e.,

$$U_{pu}(\pi) = \pi \cdot Q(\pi).$$

Show that the PU's marginal revenue curve is

$$\text{MR} = \frac{\partial U_{pu}}{\partial Q} = \pi^2 \cdot \frac{\sigma^2 G_1}{R G_0}, \quad \forall \pi \in \left[\frac{R G_0}{\sigma^2 G_1 + G_1 G_0 P_{max}}, \frac{R G_0}{\sigma^2 G_1}\right].$$

(This implies that the PU always has a positive marginal revenue, and thus he will allow the SU to transmit at the maximum power possible by setting the lowest corresponding price.)

2. We generalize the above problem to multiple SUs. That is, the PU shares its spectrum with two SUs. Let $\pi_1$ and $\pi_2$ denote the prices that the PU offers to two SUs, respectively. The SUs' utilities are

$$U_{su1}(p_1, \pi_1) = R_1 \cdot \log(1 + p_1) - p_1 \cdot \pi_1,$$
$$U_{su2}(p_2, \pi_2) = R_2 \cdot \log(1 + p_2) - p_2 \cdot \pi_2.$$

- Show the demand curves of both SUs, and the marginal revenue curve of the PU.
- Suppose that the PU has a zero marginal cost (induced by SUs' interferences). How much power will the PU sell to each SU (i.e., allow each SU to transmit), and what will be the corresponding prices?
- Suppose that the PU's marginal cost curve is given by:

$$MC = \log(1 + Q_1 + Q_2),$$

where $Q_i$ is SU $i$'s power demand. How much power will the PU sell to each SU, and what prices should he charge?

# CHAPTER 6

# Oligopoly Pricing

In Chapter 5, we considered how a single decision maker, the monopolist, chooses the price(s) to maximize the profit. In this chapter, we consider a more complicated and yet more common situation, where many self-interested individuals (including firms and consumers) make *interdependent* interactions, that is, the payoff of each individual depends not only on his own choices, but also on the choices of other individuals. Such an interaction can be analyzed by *game theory*. After introducing the basic concepts of game theory following [38, 39, 40, 41], we will look at multiple classical market competition models, including the Cournot competition based on output quantities, the Bertrand competition based on pricing, and the Hotelling model that captures the location information in the competition.

In terms of applications, we first revisit the wireless service provider competition model discussed in Chapter 4. We will study how service providers compete in the market by pricing their resources selfishly to attract customers and maximize their own revenues. In the second example, we examine how two secondary wireless service providers compete by leasing resources from spectrum owners and provide services to the same group of customers.

## 6.1 THEORY: GAME THEORY

### 6.1.1 WHAT IS A GAME?

A game is a formal representation of a situation in which a number of individuals interact with *strategic interdependence*. In other words, each individual's welfare depends not only on his own choices but also on the choices of other individuals. To describe a situation of strategic interaction, we need to define the following:

- **Players**: Who are involved in the game?

- **Rules**: What actions can players choose? How and when do they make decisions? What information do players know about each other when making decisions?

- **Outcomes**: What is the outcome of the game for each possible action combinations chosen by players?

- **Payoffs**: What are the players' preferences (i.e., utilities) over the possible outcomes?

We assume that each player is *rational* (*self-interested*), whose goal is to choose the actions that produce his most preferred outcome.[1] When facing potential uncertainty over multiple outcomes, a

---

[1]We assume that players' preference orderings are complete and transitive, like in most of the game theory literature [38, 39].

rational player chooses actions that maximize his expected utility. Under such a situation, a central issue is to identify the stable outcome(s) of the game called *equilibrium(s)*.

## 6.1.2   STRATEGIC FORM GAME

We first introduce *strategic form games* (also referred to as the normal form games). In such a game, all players make decisions simultaneously without knowing each other's choices. Thus, we only need to define the player set, the action set for each player, and the payoff (utility) function for each player. Formally,

**Definition 6.1   Strategic Form Game**   A strategic form game is a triplet $\langle \mathcal{I}, (\mathcal{S}_i)_{i \in \mathcal{I}}, (u_i)_{i \in \mathcal{I}} \rangle$ where

1. $\mathcal{I} = \{1, 2, ..., I\}$ is a finite set of players.

2. $\mathcal{S}_i$ is a set of available actions (pure strategies) for player $i \in \mathcal{I}$. We further denote by $s_i \in \mathcal{S}_i$ an action for player $i$, and by $\boldsymbol{s}_{-i} = (s_j, \forall j \neq i)$ a vector of actions for all players except $i$. We let $\mathbb{S} \triangleq \Pi_i \mathcal{S}_i$ denote the set of all action profiles. With a slight abuse of notation we write $\boldsymbol{s} = (s_i, \boldsymbol{s}_{-i}) \in \mathbb{S}$ to denote the action profile where player $i$ selects action $s_i$ and the other players use actions as described by $\boldsymbol{s}_{-i}$. We further denote $\mathbb{S}_{-i} \triangleq \Pi_{j \neq i} \mathcal{S}_j$ as the set of action profiles for all players except $i$.

3. $u_i : \mathbb{S} \to \mathbb{R}$ is the payoff (utility) function of player $i$, which maps every possible action profile in $\mathbb{S}$ to a real number, the utility.

One important concept in game theory is the *strictly dominated strategy*, which refers to a strategy that is always worse than another strategy of the same player regardless of the strategy choices of other players.

**Definition 6.2   Strictly Dominated Strategy**   A strategy $s_i \in \mathcal{S}_i$ is strictly dominated for player $i$, if there exists some $s_i' \in \mathcal{S}_i$ such that

$$u_i(s_i, \boldsymbol{s}_{-i}) < u_i(s_i', \boldsymbol{s}_{-i}), \quad \forall \boldsymbol{s}_{-i} \in \mathbb{S}_{-i} .$$

When a strategy $s_i$ is strictly dominated, it can be safely removed from player $i$'s strategy set $\mathcal{S}_i$ without changing the game outcome, as a rational player $i$ will never choose a strictly dominated strategy to maximize his payoff. This can help us to predict the outcome of some game, such as the following *Prisoner's Dilemma*.

**Example 6.3   Prisoner's Dilemma**   Two players are arrested for a crime and placed in separate rooms. The authorities try to extract a confession from them. If they both remain silent, then the authorities will not be able to press serious charges against them and they will both serve a short prison term, say two years ($u_i = -2$ for both players $i = 1, 2$), for some minor offenses. If only one of them (say, player 1) confesses, his term will be reduced to one year ($u_1 = -1$ for player 1), as a reward for him to serve as a witness against the other person, who will get a sentence of five years ($u_2 = -5$ for player 2). If they both confess, then both of them get a smaller sentence of four years ($u_i = -4$ for both players $i = 1, 2$) comparing with the worst case of five years. This game can be represented in a matrix form as follows, where each row denotes one action of player 1, each column denotes one action of player 2, and the cell indexed by row $x$ and column $y$ contains a utility pair $(a, b)$ with $a = u_1(x, y)$ and $b = u_2(x, y)$.

|  | SILENT | CONFESS |
|---|---|---|
| SILENT | $(-2, -2)$ | $(-5, -1)$ |
| CONFESS | $(-1, -5)$ | $(-4, -4)$ |

Now we show that "SILENT" is a strictly dominated strategy for both players. Let us first look at player 1 (the row player). When player 2 chooses "SILENT" (the first column), then player 1 obtains a worse payoff of -2 when he chooses "SILENT," comparing with the payoff of -1 if he chooses "CONFESS." When player 2 chooses "CONFESS" (the second column), then player 1 obtains a worse payoff of -5 when he chooses "SILENT," comparing with the payoff of -4 if he chooses "CONFESS." This means that for player 1, the strategy "SILENT" is always worse than "CONFESS," and can be eliminated from his strategy set. Since this game is symmetric, the same conclusion is true for player 2. Hence, the unique game result is (CONFESS, CONFESS), and the payoffs of both players are $(-4, -4)$.

However, most of the time we cannot predict a game's outcome by eliminating strictly dominated strategies. Next we introduce the more general method of predicting the game outcome by looking at the *Best Response Correspondence*.

**Definition 6.4  Best Response Correspondence**   For each player $i$, the best response correspondence $B_i(s_{-i}) : \mathbb{S}_{-i} \to \mathcal{S}_i$ is a mapping from the set $\mathbb{S}_{-i}$ into $\mathcal{S}_i$ such that

$$B_i(s_{-i}) = \{s_i \in \mathcal{S}_i \mid u_i(s_i, s_{-i}) \geq u_i(s_i', s_{-i}), \forall s_i' \in \mathcal{S}_i\}.$$

Let's use the best response correspondence concept to predict the result of the following *Stag Hunt* game.

**Example 6.5  Stag Hunt**   Two hunters decide to hunt together in a forest, and each of them chooses one animal to hunt: stag or hare. Hunting a stag is challenging and requires cooperation between the hunters to succeed. Hunting a hare is easy and can be done by a single hunter. When both hunters choose to hunt a stag, each of them will get a payoff of 10 (pounds of stag meat). When hunter 1 hunts a stag but hunter 2 hunts a hare, then hunter 1 gets nothing due to the lack of cooperation, and hunter 2 gets a payoff of 2 (pounds of hare meat). The situation is similar when hunter 1 hunts a hare and hunter 2 hunts a stag. Finally, when both hunters hunt hares, then each of them will get a payoff of 2 as there are enough hares around in the forest. This game can be represented in the matrix form as follows.

|  | STAG | HARE |
|---|---|---|
| STAG | $(10, 10)$ | $(0, 2)$ |
| HARE | $(2, 0)$ | $(2, 2)$ |

One can easily verify that there is no strictly dominated strategy in this game, as choosing "HARE" is worst than "STAG" for hunter 1 (row player) when hunter 2 chooses "STAG" (the first column), but choosing "STAG" is worst than "HARE" for hunter 1 when hunter 2 chooses "HARE" (the second column). However, the above analysis reveals that the hunter 1's best response functions are $B_1(STAG) = STAG$ and $B_1(HARE) = HARE$. As the game is symmetric, we can similarly show that $B_2(STAG) = STAG$ and $B_2(HARE) = HARE$. This shows that there are two strategy profiles, (STAG, STAG) and (HARE, HARE), they are mutual best responses of both players. For example, if hunter 2 chooses "STAG," then it is hunter 1's best response of choosing "STAG," and vice versa. Hence, (STAG, STAG) is a robust predication of the game result. So is (HARE, HARE). In fact, this naturally leads to the concept of *Nash Equilibrium* defined next.

### 6.1.3   NASH EQUILIBRIUM

**Definition 6.6   Pure Strategy Nash Equilibrium**   A pure strategy Nash Equilibrium of a strategic form game $\langle \mathcal{I}, (\mathcal{S}_i)_{i \in \mathcal{I}}, (u_i)_{i \in \mathcal{I}} \rangle$ is a strategy profile $s^* \in \mathbb{S}$ such that for each player $i \in \mathcal{I}$ the following condition holds

$$u_i(s_i^*, s_{-i}^*) \geq u_i(s_i', s_{-i}^*), \quad \forall s_i' \in \mathcal{S}_i.$$

The above definition can be restated in terms of best-response correspondences:

**Definition 6.7   Pure Strategy Nash Equilibrium-Restated**   A strategy profile $s^* \in \mathbb{S}$ is a Nash Equilibrium of a strategic form game $\langle \mathcal{I}, (\mathcal{S}_i)_{i \in \mathcal{I}}, (u_i)_{i \in \mathcal{I}} \rangle$ if and only if

$$s_i^* \in B_i(s_{-i}^*), \quad \forall i \in \mathcal{I},$$

where $B_i(\cdot)$ is the best response correspondence defined in Definition 6.4.

Hence, we know that the *Stag Hunt* game has two pure strategy Nash equilibria: (STAG, STAG) and (HARE, HARE). Moreover, we can verify that the *Prisoner's Dilemma* has a unique pure strategy Nash equilibrium: (CONFESS, CONFESS).

We note that a pure strategy Nash Equilibrium may not be a Pareto optimal solution. That is, there may be other strategy profiles under which *all* players achieve higher utilities than under a Nash Equilibrium. For example, in the *Prisoner's Dilemma*, choosing (SILENT, SILENT) will lead to payoffs of $(-2, -2)$, which are better than the payoffs of $(-4, -4)$ under the Nash equilibrium (CONFESS, CONFESS). Such a loss is due to the selfish nature of the players.

It is worth noting that *not* every game possesses a pure strategy Nash Equilibrium. To see this, let us consider the following *Matching Pennies* Game.

**Example 6.8  Matching Pennies**    The game is played between two players. Each player has a penny and turns his penny to "HEADS" or "TAILS" secretly and simultaneously with the other player. If the pennies match (both heads or both tails), player 1 keeps both pennies, so wins one from player 2 ($u_1 = 1$ for player 1, $u_2 = -1$ for player 2). If the pennies do not match (one heads and one tails), player 2 keeps both pennies, so receives one from player 1 ($u_1 = -1$ for player 1, $u_2 = 1$ for player 2). This is an example of a *zero-sum game*, where one player's gain is exactly the other player's loss. This game can be represented by the following matrix.

|        | HEADS     | TAILS     |
|--------|-----------|-----------|
| HEADS  | $(1, -1)$ | $(-1, 1)$ |
| TAILS  | $(-1, 1)$ | $(1, -1)$ |

It is easy to verify that player 1's best response is "HEADS" if player 2 selects "HEADS," and "TAILS" if player 2 selects "TAILS." However, player 2's best response functions are exactly the opposite. Hence, there is no strategy profile that corresponds mutual best responses of both players, and hence there is no pure strategy Nash Equilibrium.

Another way to understand the lack of pure strategy equilibrium in the *Matching Pennies* game is to consider the following cyclic best response behavior. When player 1 chooses "HEADS," player 2 will choose "TAILS" as his best response. In response to this, player 1 will choose "TAILS," which makes player 2 choose "HEADS." Because of this, player 1 will switch to "HEADS," hence enters a loop.

When a game does not have a pure strategy Nash Equilibrium, it is natural to ask what kind of outcome will emerge as an "equilibrium?" If we allow the players to *randomize* over their actions, then we can show that the game will reach a *mixed strategy Nash Equilibrium*.

We need some new notations to explain this new equilibrium concept. Let $\sigma_i$ denote a mixed strategy for player $i$, which is a probability distribution function (or probability mass function over the finite set $\mathcal{S}_i$) over all pure strategies in set $\mathcal{S}_i$. For example, in the *Matching Pennies* game, $\sigma_1 = (0.4, 0.6)$ is a mixed strategy for player 1, which means that player 1 picks "HEADS" with probability 0.4 and "TAILS" with probability 0.6. Let $\Sigma_i$ denote the set of all mixed strategies of player $i$, i.e., all probability distributions over $\mathcal{S}_i$. Let $\boldsymbol{\sigma} = (\sigma_i)_{i \in \mathcal{I}} \in \Sigma$ denote a mixed strategy profile for all players, where $\Sigma = \Pi_i \Sigma_i$ is the set of all mixed strategy profiles. Furthermore, let $\boldsymbol{\sigma}_{-i} = (\sigma_j, \forall j \neq i)$ denote a mixed strategy profile for all players except $i$, and $\Sigma_{-i} = \Pi_{j \neq i} \Sigma_j$ denote the set of mixed strategy profile for all players except $i$.

Each player $i$'s payoff under a mixed strategy profile $\boldsymbol{\sigma}$ is given by the expected value of pure strategy payoffs under the distribution $\sigma$. More precisely, we have

$$u_i(\boldsymbol{\sigma}) = \sum_{\boldsymbol{s} \in \mathbf{S}} \left( \Pi_{j=1}^I \sigma_j(s_j) \right) \cdot u_i(\boldsymbol{s}), \qquad (6.1)$$

where $\boldsymbol{s} = (s_j, \forall j \in \mathcal{I})$ is a pure strategy profile, and $\Pi_{j=1}^I \sigma_j(s_j)$ is the probability of choosing a particular pure strategy profile $\boldsymbol{s}$.

Based on above, the mixed strategy Nash Equilibrium is defined as follows [38, 39].

**Definition 6.9   Mixed Strategy Nash Equilibrium**   A mixed strategy profile $\boldsymbol{\sigma}^*$ is a mixed strategy Nash Equilibrium if for every player $i \in \mathcal{I}$,

$$u_i(\sigma_i^*, \boldsymbol{\sigma}_{-i}^*) \geq u_i(\sigma_i', \boldsymbol{\sigma}_{-i}^*), \quad \forall \sigma_i' \in \Sigma_i.$$

Let $\text{supp}(\sigma_i)$ denote the support of $\sigma_i$, defined as the set $\text{supp}(\sigma_i) \triangleq \{s_i \in \mathcal{S}_i \mid \sigma_i(s_i) > 0\}$, that is, the support of $\sigma_i$ is the set of pure strategies which are assigned positive probabilities. We have the following useful characterization of a mixed strategy Nash Equilibrium.

**Proposition 6.10**   *A mixed strategy profile $\boldsymbol{\sigma}^*$ is a mixed strategy Nash Equilibrium if and only if for every player $i \in \mathcal{I}$, the following two conditions hold:*

1. *Every chosen action is equally good, that is, the expected payoff given $\boldsymbol{\sigma}_{-i}^*$ of every $s_i \in \text{supp}(\sigma_i)$ is the same.*

2. *Every non-chosen action is not good enough, that is, the expected payoff given $\boldsymbol{\sigma}_{-i}^*$ of every $s_i \notin \text{supp}(\sigma_i)$ must be no larger than the expected payoff of $s_i \in \text{supp}(\sigma_i)$.*

Intuitively, Proposition 6.10 states that for a player $i$, every action in the support of a mixed strategy Nash Equilibrium is a best response to $\boldsymbol{\sigma}_{-i}^*$. This proposition follows from the fact that if the strategies in the support have different payoffs, then it would be better to just take the pure strategy with the highest expected payoff. This would contradict the assumption that $\boldsymbol{\sigma}^*$ is a Nash Equilibrium. Using the same argument, it follows that the pure strategies which are not in the support must have lower (or equal) expected payoffs compared to the ones in the support set.

For the *Matching Pennies* game, we can show that $\boldsymbol{\sigma}^* = (\sigma_1^*, \sigma_2^*)$ with $\sigma_i^* = (0.5, 0.5)$, $i = 1, 2$, is the unique mixed strategy Nash Equilibrium. To see this, we can verify that given player 2's equilibrium strategy $\sigma_2^* = (0.5, 0.5)$, player 1 achieves the same best expected payoff with both pure strategies.

An important problem is to determine when a strategic form game possesses a (pure or mixed) strategy Nash equilibrium. Notice that a pure strategy Nash Equilibrium is a special case of the mixed strategy Nash Equilibrium. The most important results regarding this are listed in the following theorems.

**Theorem 6.11   Existence (Nash 1950)**   *Any finite strategic game, i.e., a game that has a finite number of players and each player has a finite number of action choices, has at least one mixed strategy Nash Equilibrium.*

**Theorem 6.12   Existence (Debreu-Fan-Glicksburg 1952)**   *The strategic form game $\langle \mathcal{I}, (\mathcal{S}_i)_{i \in \mathcal{I}}, (u_i)_{i \in \mathcal{I}} \rangle$ has a pure Nash equilibrium, if for each player $i \in \mathcal{I}$ the following conditions hold:*

*1. $\mathcal{S}_i$ is a non-empty, convex, and compact subset of a finite-dimensional Euclidean space.*

*2. $u_i(\boldsymbol{s})$ is continuous in $\boldsymbol{s}$ and quasi-concave[2] in $s_i$.*

Both theorems can be proved by the Kakutani fixed point theorem (see [42]). Theorem 6.11 guarantees the existence of Nash equilibrium in games like *Prisoner's Dilemma*, *Stag Hunt*, and *Matching Pennies*. Theorem 6.12 helps us to understand the existence of a pure Nash equilibrium in a strategic form game where players have continuous strategy sets (e.g., the Cournot Competition Game in Section 6.2.1).

## 6.1.4   EXTENSIVE FORM GAME

We have studied the strategic form games which are used to model one-shot games, in which each player chooses his action once and all players act simultaneously. In this section, we further study the *extensive form games*, where players engage in sequential decision making [38, 39]. Our focus will be on multi-stage games with observed actions where:

1. All previous actions (called history) are observed, i.e., each player is perfectly informed of all previous events.

2. Some players may move simultaneously within the same stage.

---

[2]A function is quasi-concave if its negative is quasi-convex. A quasi-convex function is a real-valued function defined on an interval or on a convex subset of a real vector space, such that the inverse image of any set of the form $(-\infty, a)$ is a convex set.

Extensive form games can be conveniently represented by *tree* diagrams, such as the following *Market Entry* game.

**Example 6.13   Market Entry**   There are two players (firms). Player 1, the challenger, can choose to enter the market (I) or stay out (O). Player 2, the monopolist, after observing the action of the challenger, chooses to accommodate (A) or fight (F). The detailed process is shown in Figure 6.1. Note that when player 1 chooses "Out," there will be no difference for the player 2 to choose "Fight" or "Accord." At each of the four leaf nodes, we show the payoff pair of both players. For example, when player 1 chooses "IN" and player 2 chooses "ACCORD," then player 1 gets a payoff of 2 and player 2 gets a payoff of 1.

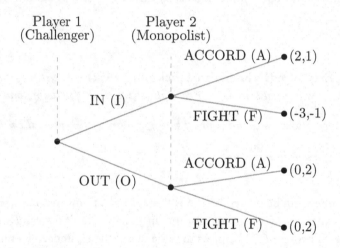

**Figure 6.1:** Market Entry Game.

An extensive form game can be formally defined as follows [38, 39, 43].

**Definition 6.14   Extensive Form Game**   An extensive form game consists of four main elements:

1. A set of players $\mathcal{I} = \{1, 2, ..., I\}$.

2. Histories: A set $\mathcal{H}$ of sequences which can be finite or infinite, defined by

$$\begin{cases} h^0 = \emptyset & \text{initial history} \\ h^1 = (s^0) & \text{history after stage 0} \\ h^2 = (s^0, s^1) & \text{history after stage 1} \\ ... & ... \\ h^{k+1} = (s^0, ..., s^k) & \text{history after stage } k \end{cases}$$

where $s^t = (s_i^t, \forall i \in \mathcal{I})$ is the action profile at stage $t$.

If the game has a finite number $(K + 1)$ of stages (i.e., from stage 0 to stage $K$), then it is a finite horizon game. Let $\mathcal{H}^k = \{h^k\}$ be the set of all possible histories after stage $k - 1$ (i.e., at stage $k$). Then $\mathcal{H}^{K+1} = \{h^{K+1}\}$ is the set of all possible terminal histories (after stage $K$), and $\mathcal{H} = \bigcup_{k=0}^{K+1} \mathcal{H}^k$ is the set of all possible histories. Consider the *Market Entry* game in Figure 6.1, we have $\mathcal{H}^1 = \{I, O\}$ and $\mathcal{H}^2 = \{(I, A), (I, F), (O, A), (O, F)\}$.

3. Each pure strategy for player $i$ is defined as a contingency plan for every possible history. Let $\mathcal{S}_i(h^k)$ be the set of actions available to player $i$ under history $h^k$, and $\mathcal{S}_i(\mathcal{H}^k) = \bigcup_{h^k \in \mathcal{H}^k} \mathcal{S}_i(h^k)$ be the set of actions available to player $i$ under all possible histories at stage $k$. Let $a_i^k : \mathcal{H}^k \to \mathcal{S}_i(\mathcal{H}^k)$ be a mapping from $\mathcal{H}^k$ to $\mathcal{S}_i(\mathcal{H}^k)$ such that $a_i^k(h^k) \in \mathcal{S}_i(h^k)$. Then a pure strategy of player $i$ is a sequence $s_i = \{a_i^k\}_{k=0}^K$. The collection of all such sequences $s_i$ forms the set of strategies available to player $i$. The path of strategy profile $s$ includes $s^0 = a^0(h^0)$, $s^1 = a^1(s^0)$, $s^2 = a^2(s^0, s^1)$, and so on, where $a^k(\cdot) = (a_i^k(\cdot), \forall i \in \mathcal{I})$.

4. Preferences are defined on the outcome of the game $\mathcal{H}^{K+1}$ (after stage $K$). We can represent the preferences of player $i$ by a utility function $u_i : \mathcal{H}^{K+1} \to \mathbb{R}$. As the strategy profile $s$ determines the path $(s^0, ..., s^k)$, and hence $h^{K+1}$, we will denote the payoff to player $i$ under strategy profile $s$ as $u_i(s)$.

We want to emphasize that in an extensive form game, a strategy specifies the action the player chooses for *every* possible history. This is very different from the strategic form game, where "pure strategy" and "action" have the same meaning. Consider the *Market Entry* game in Figure 6.1. Player 1 moves in the first stage (i.e., stage 0) and player 2 moves in the second stage (i.e., stage 1). The strategy of player 1 is the function $a_1^0 : \mathcal{H}^0 = \emptyset \to \mathcal{S}_1 = \{I, O\}$. The strategy of player 2 is the function $a_2^1 : \mathcal{H}^1 = \{I, O\} \to \mathcal{S}_2(\mathcal{H}^1)$. There are four possible strategies for player 2, which we can represent as AA, AF, FA, and FF, each corresponding to a contingency plan of player 2 for every possible history in $\mathcal{H}^1 = \{I, O\}$. That is, the strategy AA means that player 2 will select "ACCORD" under both histories in $\mathcal{H}^1 = \{I, O\}$; the strategy FA means that player 2 will select "FIGHT" under history $h^1 = \{I\}$ and "ACCORD" under history $h^1 = \{O\}$. If the strategy profile is (I, AF) or (I, AA), then the outcome will be (player 1 chooses "IN," player 2 chooses "ACCORD"). On the other hand, if the strategy profile is (O, FA) or (O, AA), then the outcome will be (player 1 chooses "OUT," player 2 chooses "ACCORD").

Based on above discussions, one might want to represent the extensive form game in the corresponding strategic form, and solve the equilibria using the methods introduced in Section 6.1.3. The following matrix shows the strategic form of the *Market Entry* game. Each row denotes a strategy of player 1, and each column denotes a strategy of player 2.

By checking the best responses of both players, one can conclude that there are four pure strategy Nash Equilibria in this game: (I, AA), (I, AF), (O, FA), and (O, FF). However, further thinking reveals that the two equilibria (O, FA) and (O, FF) are problematic, as both of them rely

|   | AA | AF | FA | FF |
|---|---|---|---|---|
| I | (2, 1) | (2, 1) | (−3, −1) | (−3, −1) |
| O | (0, 2) | (0, 2) | (0, 2) | (0, 2) |

on the *empty threat* that player 2 will choose "FIGHT" when player 1 chooses "IN." To see why this will not happen, we notice that once player 1 chooses "IN," then player 2 will definitely choose "ACCORD" and get a payoff of 1 (as "FIGHT" leads to a worse payoff of -1). This will eliminate player 2's two strategies: "FA" and "FF." Hence, the only "reasonable" Nash equilibria will be (I,AA) and (I,AF).

Next we make the above discussions formal by introducing the concept of subframe perfect equilibrium.

### 6.1.5   SUBGAME PERFECT EQUILIBRIUM

The *subgame perfect equilibrium* (SPE) requires the strategy of each player to be optimal not only at the start of the game but also after every history [44]. Let $h^k$ denote a history at stage $k$. We define $G(h^k)$ as the game from $h^k$ on with

- Histories: $h^{K+1} = (h^k, s^k, ..., s^K)$.

- Strategies: $s_{i|h^k}$ is the restriction of $s_i$ to histories in $G(h^k)$.

- Payoffs: $u_i(s_i, s_{-i}|h^k)$ is the payoff of player $i$ after histories in $G(h^k)$.

Such a game is referred to as the *subgame* from history $h^k$. Then a subgame perfect equilibrium is defined as follows:

**Definition 6.15   Subgame Perfect Equilibrium**   A strategy profile $s^*$ is a subgame perfect equilibrium for an extensive form game if for every history $h^k$, the restriction $s^*_{i|h^k}$ is an Nash equilibrium of the subgame $G(h^k)$.

For finite horizon games, the subgame perfect equilibria can be derived using *backward induction*. Let us solve the SPE of the *Market Entry* game using this method.

We first look at the two subgames from history $h^1 = \{I\}$ and $h^1 = \{O\}$, which concern the decision of player 2 in the last stage. In the first subgame $G$ (I) as in Figure 6.2, it is clear that player 2 will choose "ACCORD" to maximize his payoff (as 1 is better than -1), and hence we can eliminate "FIGHT." In the second subgame $G$ (O) as in Figure 6.3, player 2 is indifferent from choosing "ACCORD" or "FIGHT," hence we cannot eliminate any action.

Now we proceed to the remaining subgame, which concerns player 1's decision in the initial stage as in Figure 6.4. Player 1 now faces two possible payoffs: he will get 2 if he chooses "IN," and he will get 0 if he chooses "OUT." So clearly he will choose "IN." Hence, the SPEs are (I, AA) and (I, AF). The final *equilibrium path* would be (player 1 chooses "IN," player 2 chooses "ACCORD"), and the *equilibrium payoffs* are (2, 1).

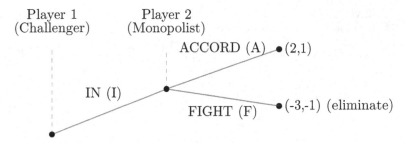

Figure 6.2: Market Entry Game: subgame $G$ (I).

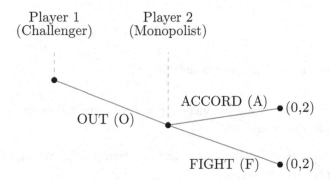

Figure 6.3: Market Entry Game: subgame $G$ (O).

## 6.2   THEORY: OLIGOPOLY

Now we consider three classical strategic form game formulations for competitions among multiple entities (also called Oligopoly) [45]: the Cournot model, the Bertrand model, and the Hotelling model. We use these models to illustrate: (a) the translation of an informal problem statement into a strategic form representation of a game; and (b) the analysis of Nash equilibrium when a player can choose his strategy from a continuous set.

### 6.2.1   THE COURNOT MODEL

The Cournot model is an economic model used to describe interactions among firms that compete on the amount of output they will produce, which they decide independently of each other simultaneously [46]. It is named after Antoine Augustin Cournot (1801–1877). A Cournot model usually has the following key features:

• There are at least two firms producing homogeneous (undifferentiated) products;

• Firms do not cooperate, i.e., there is no collusion;

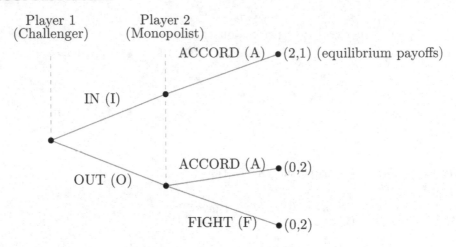

**Figure 6.4:** Market Entry Game: subgame $G(\emptyset)$.

- Firms compete by setting production quantities simultaneously. The total output quantity affects the market price;

- The firms are economically rational and act strategically, seeking to maximize profits given their competitors' decisions.

For simplicity, we consider a Cournot model between two firms, $\mathcal{I} = \{1, 2\}$. Each firm $i$ decides his output quantity $q_i$, under a fixed unit producing cost $c_i$. The market-clearing price is a decreasing function of the total quantity $Q = q_1 + q_2$, denoted by $P(Q)$. In such a competition model, what is the best quantity choice of each firm?

We first translate the problem into a strategic form game—*Cournot Game*. Recall from Definition 6.1, we have the following strategic form representation of this Cournot game:

- The set of players is $\mathcal{I} = \{1, 2\}$,

- The strategy set available to each player $i \in \mathcal{I}$ is the set of all nonnegative real numbers, i.e., $q_i \in [0, \infty)$,

- The payoff received by each player $i$ is a function of both players' strategies, defined by $\Pi_i(q_i, q_{-i}) = q_i \cdot P(Q) - c_i \cdot q_i$. The first term denotes the player $i$'s revenue from selling $q_i$ units of products at a market-clearing price $P(Q)$, and the second term denotes the player $i$'s production cost.

Next we identify the Nash equilibrium of this Cournot game. Given player 2's strategy $q_2$, player 1's payoff (profit) is a function of his quantity $q_1$,

$$\Pi_1(q_1, q_2) = q_1 \cdot P(q_1 + q_2) - c_1 \cdot q_1.$$

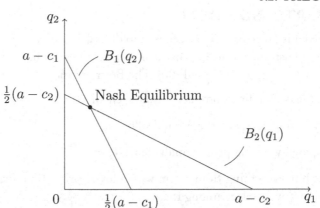

**Figure 6.5:** The Cournot Game.

The best response of player 1 is the strategy $q_1$ which maximizes this function, given player 2's strategy $q_2$. The optimal strategy for player 1 needs to satisfy the following first-order condition (ignoring boundary conditions):

$$q_1 \cdot P'(q_1 + q_2) + P(q_1 + q_2) - c_1 = 0.$$

As an example, we assume that $P(q_1 + q_2) = a - q_1 - q_2$. Then the best response of player 1 is

$$q_1^* = B_1(q_2) = \frac{a - q_2 - c_1}{2},$$

which is a function of player 2's strategy $q_2$.

Similarly, given player 1's strategy $q_1$, the optimal strategy for player 2 or the best response of player 2 is

$$q_2^* = B_2(q_1) = \frac{a - q_1 - c_2}{2},$$

which is a function of player 1's strategy $q_1$.

Recall from Definition 6.6, a strategy profile $(q_1^*, q_2^*)$ is an Nash equilibrium if every player's strategy is the best response to others' strategies, that is, $q_1^* = B_1(q_2^*)$ and $q_2^* = B_2(q_1^*)$. This directly leads to the following pure strategy Nash equilibrium:

$$q_1^* = \frac{a + c_1 + c_2}{3} - c_1, \quad q_2^* = \frac{a + c_1 + c_2}{3} - c_2.$$

Here we assume that $a$ is large enough such that $q_1^* > 0$ and $q_2^* > 0$.

Figure 6.5 illustrates both players' best response functions and the Nash equilibrium. Geometrically, the Nash equilibrium is the intersection of both players' best response curves. For more information about variations of Cournot games, please refer to [38, 39, 46].

## 6.2.2 THE BERTRAND MODEL

The Bertrand model is an economic model used to describe interactions among firms (sellers) that set prices and their customers (buyers) that choose quantities at that price [46]. It is named after Joseph Louis Francois Bertrand (1822–1900). The Bertrand model has the following key features:

- There are at least two firms producing homogeneous (undifferentiated) products;

- Firms do not cooperate, i.e., there is no collusion;

- Firms compete by setting prices simultaneously;

- Consumers buy everything from a firm with a lower price. If all firms charge the same price, consumers randomly select among them.

- The firms are economically rational and act strategically, seeking to maximize profits given their competitors' decisions.

Similarly, we consider a Bertrand model between two firms, $\mathcal{I} = \{1, 2\}$. Each firm $i$ chooses the price $p_i$, rather than quantity as in the Cournot model. Consumers buy from the firm with a lower price, and the total consumer demand is a decreasing function of the market price, denoted by $D(\min\{p_1, p_2\})$. In such a competition model, what is the best price choice of each firm? It is important to note that the Bertrand model is a different game than the Cournot model: the strategy spaces are different, the payoff functions are different, and (as will be shown later) the market outcomes in the Nash equilibria of the two models are different.

The strategic form representation for this (two firms) Bertrand model, called *Bertrand Game*, is as follows:

- The set of players is $\mathcal{I} = \{1, 2\}$,

- The strategy set available to each player $i \in \mathcal{I}$ is the set of all nonnegative real numbers, i.e., $p_i \in [0, \infty)$,

- The payoff (profit) received by each player $i$ is a function of both players' strategies, defined by $\Pi_i(p_i, p_{-i}) = (p_i - c_i) \cdot D_i(p_1, p_2)$, where $c_i$ is the unit producing cost and $D_i(p_1, p_2)$ is the consumers' demand to player $i$.

Obviously, if player $i$'s price is lower than that of the other player (denoted by $-i$), then he gets the total consumer demand $D(P)$; and if two players' prices are the same, each player gets half of the total consumer demand $D(P)$. That is, $D_i(p_1, p_2) = D(p_i)$ if $p_i < p_{-i}$; $D_i(p_1, p_2) = 0$ if $p_i > p_{-i}$; and $D_i(p_1, p_2) = D(p_i)/2$ if $p_i = p_{-i}$.

Next we identify the (unique) Nash equilibrium of this Bertrand game. Given player 2's strategy $p_2$, player 1's payoff is a function of his price $p_1$,

$$\Pi_1(p_1, p_2) = \begin{cases} (p_1 - c_i) \cdot D(p_1) & \text{if } p_1 < p_2 \\ 0 & \text{if } p_1 > p_2 \\ (p_1 - c_i) \cdot D(p_1)/2 & \text{if } p_1 = p_2 \end{cases}$$

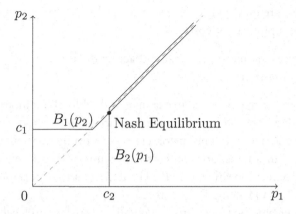

**Figure 6.6:** The Bertrand Game.

Thus, given player 2's strategy $p_2$, the optimal strategy for player 1 or the best response of player 1 is to select a price $p_1$ slightly lower than $p_2$, under the constraint that $p_1 \geq c_1$.

Similarly, given player 1's strategy $p_1$, the optimal strategy for player 2 or the best response of player 2 is to select a price $p_2$ slightly lower than $p_1$, under the constraint that $p_2 \geq c_2$. Thus, for any strategy profile $(p_1, p_2)$, both players will gradually decrease their prices, until one player gets to his lowest acceptable price, i.e., his producing cost. Therefore, the Nash equilibrium is given by

$$
\begin{cases}
p_1^* = [c_2]^-, & p_2^* \in [c_2, \infty) & \text{if } c_1 < c_2 \\
p_1^* \in [c_1, \infty), & p_2^* = [c_1]^- & \text{if } c_1 > c_2 \\
p_1^* = p_2^* = c & & \text{if } c_1 = c_2 = c
\end{cases}
$$

where $[x]^-$ denotes the value slightly lower than $x$. The above Nash equilibrium implies that the lower producing cost firm will extract all the consumer demand, by setting a price slightly lower than the other firm's producing cost. This is geometrically illustrated in Figure 6.6. Note that the above classic Bertrand model assumes firms compete purely on price, ignoring non-price competition. In a more general case, firms can differentiate their products and charge a higher price. For detailed information, please refer to [38, 39, 46].

### 6.2.3   THE HOTELLING MODEL

The Hotelling model is an economic model used to study the effect of locations on the competition among two or more firms [46]. It is named after Harold Hotelling (1895–1973). The Hotelling model has the following key features:

- There are two firms selling the same good. The firms have different locations, which are represented by two points in the interval [0, 1].

- The customers are uniformly distributed along the interval. Customers incur a transportation cost as well as a purchasing cost.

- The firms are economically rational and act strategically, seeking to maximize profits given their competitors' decisions.

We take the following model as an example of the Hotelling model. Consider a one mile long beach on a hot summer day. There are two identical icecream shops on both ends of the beach: store 1 at $x = 0$ and store 2 at $x = 1$. The customers are uniformly distributed with density 1 along this beach. Customers incur a transportation cost $w$ per unit of length (e.g., the value of time spent in travel). Thus, a customer at location $x \in [0, 1]$ will incur a transportation cost of $w \cdot x$ when going to store 1 and $w \cdot (1 - x)$ when going to store 2.

Each customer buys one icecream and obtains a satisfaction level of $\bar{s}$, which is large enough such that all customers want to purchase one icecream from one of the stores. Each store $i \in \{1, 2\}$ chooses a unit price $p_i$. A customer will choose a store that has the less generalized cost, i.e., price plus transportation cost. Each store wants to choose the price to maximize his own profit, by taking the unit cost into consideration.

The strategic form representation for this Hotelling model, called *Hotelling Game*, is as follows:

- The set of players is $\mathcal{I} = \{1, 2\}$,

- The strategy set available to each player $i \in \mathcal{I}$ is the set of all nonnegative real numbers, i.e., $p_i \in [0, \infty)$,

- The payoff received by each player $i$ is a function of both players' strategies, defined by $\Pi_i(p_i, p_{-i}) = (p_i - c_i) \cdot D_i(p_1, p_2)$, where $c_i$ is the unit producing cost and $D_i(p_1, p_2)$ is the ratio of consumers coming to player $i$ (which will be analyzed later).

Next we derive the Nash equilibrium of this game. First, let us compute the location of the customer who is indifferent of choosing either store, $x = l(p_1, p_2)$, where $x$ is given by equating the generalized costs,

$$p_1 + w \cdot x = p_2 + w \cdot (1 - x).$$

Thus, the players' respective demand ratios are

$$D_1(p_1, p_2) = l(p_1, p_2) = \frac{p_2 - p_1 + w}{2w}$$

and

$$D_2(p_1, p_2) = 1 - l(p_1, p_2) = \frac{p_1 - p_2 + w}{2w}.$$

Given player 2's price $p_2$, the profit of player 1 is given by

$$\Pi_1(p_1, p_2) = (p_1 - c_1) \cdot \frac{p_2 - p_1 + w}{2w}.$$

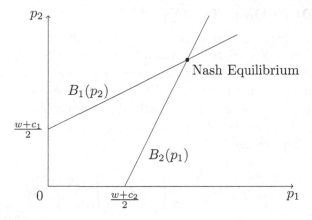

**Figure 6.7:** The Hotelling Game.

Thus, the optimal strategy for player 1 or the best response of player 1 is given by the first order condition, i.e.,

$$p_1^* = B_1(p_2) = \frac{p_2 + w + c_1}{2}.$$

Similarly, given player 1's price $p_1$, the best response of player 2 is given by

$$p_2^* = B_2(p_1) = \frac{p_1 + w + c_2}{2}.$$

The Nash equilibrium of the Hotelling game is given by $p_1^* = B_1(p_2^*)$ and $p_2^* = B_2(p_1^*)$, i.e.,

$$p_1^* = \frac{3w + c_1 + c_2}{3} + \frac{c_1}{3}, \quad p_2^* = \frac{3w + c_1 + c_2}{3} + \frac{c_2}{3}.$$

Figure 6.7 illustrates both players' best response functions and the Nash equilibrium. Geometrically, the Nash equilibrium is the intersection of both players' best response curves. Note that the above classic Hotelling model assumes firms compete purely on price with fixed locations. In a more general case, firms can choose different locations so as to attract more consumers. For detailed information, please refer to [38, 39, 46].

## 6.3 APPLICATION I: WIRELESS SERVICE PROVIDER COMPETITION REVISITED

Here we revisit the wireless service provider competition model in Section 4.3. In Section 4.3, we solve the social welfare optimization problem, assuming that all providers are under the control of the same entity (e.g., regulator). In this section, we look at the market competition case, where each provider determines its own price to maximize its own revenue. This can be modeled as a multi-leader-follower provider competition game.

### 6.3.1  PROVIDER COMPETITION GAME

We consider a set $\mathcal{J} = \{1, \ldots, J\}$ of service providers and a set $\mathcal{I} = \{1, \ldots, I\}$ of users. The provider competition game consists of two stages. In the first stage, providers announce prices $\boldsymbol{p} = (p_j, \forall j \in \mathcal{J})$, where $p_j$ is the unit resource price announced by provider $j$. In the second stage, each user $i \in \mathcal{I}$ chooses a demand vector $\boldsymbol{q}_i = (q_{ij}, \forall j \in \mathcal{J})$, where $q_{ij}$ represents the demand from user $i$ to provider $j$. We use $\boldsymbol{q} = (\boldsymbol{q}_i, \forall i \in \mathcal{I})$ to denote the demand vector of all users.

In the second stage where prices $\boldsymbol{p}$ are known, the goal of user $i$ is to choose $\boldsymbol{q}_i$ to maximize its payoff (utility minus payment):

$$v_i(\boldsymbol{q}_i, \boldsymbol{p}) = u_i\left(\sum_{j=1}^{J} q_{ij} c_{ij}\right) - \sum_{j=1}^{J} p_j q_{ij}, \tag{6.2}$$

where $c_{ij}$ is the *channel quality offset* for the channel between user $i$ and the base station of provider $j$ (see Example 4.11 and Assumption 4.13), and $u_i$ is an increasing and strictly concave utility function. In the first stage, a provider $j$ chooses price $p_j$ to maximize its revenue $p_j \sum_{i \in \mathcal{I}} q_{ij}$ subject to the resource constraint $\sum_{i \in \mathcal{I}} q_{ij} \leq Q_j$, while taking into account the effect of all prices on the user demands in the second stage.

### 6.3.2  ANALYSIS OF THE TWO-STAGE GAME

Next we will show that there exists a unique subgame perfect equilibrium (SPE) of the two-stage game, and such an equilibrium corresponds to the unique social optimal solution of SWO and the associated Lagrange multipliers discussed in Section 4.3.2 under fairly mild technical assumptions. The idea is to show that the optimal Lagrange multipliers coincide with the prices announced by the providers at the SPE.

In this game, a price demand tuple $(\boldsymbol{p}^*, \boldsymbol{q}^*(\boldsymbol{p}^*))$ is an SPE if no player has an incentive to deviate unilaterally at any stage of the game. In particular, each user $i \in \mathcal{I}$ maximizes its payoff by choosing the demand $\boldsymbol{q}_i^*(\boldsymbol{p}^*)$ given prices $\boldsymbol{p}^*$. Each provider $j \in \mathcal{J}$ maximizes its revenue by choosing the price $p_j^*$ given other providers' prices $\boldsymbol{p}_{-j}^* = (p_k^*, \forall k \neq j)$ and the user demands $\boldsymbol{q}^*(\boldsymbol{p}^*)$.

We will compute the SPE using backward induction. In Stage II, we will compute the best response of the users $\boldsymbol{q}^*(\boldsymbol{p})$ as a function of any given price vector $\boldsymbol{p}$. Then in Stage I, we will compute the equilibrium prices $\boldsymbol{p}^*$.

**Equilibrium strategy of the users in Stage II**

Consider users facing prices $\boldsymbol{p}$ in the second stage. Each user $i \in \mathcal{I}$ solves a user payoff maximization (UPM) problem:

$$\textbf{UPM} : \max_{\boldsymbol{q}_i \geq 0} \left( u_i\left(\sum_{j=1}^{J} q_{ij} c_{ij}\right) - \sum_{j=1}^{J} p_j q_{ij} \right). \tag{6.3}$$

We can show that the optimal solution of Problem UPM is unique in terms of the effective resource $x_i$ as defined in Section 4.3.2.

**Lemma 6.16**    *For each user $i \in \mathcal{I}$, there exists a unique nonnegative value $x_i^*$, such that $\sum_{j=1}^{¥} c_{ij} q_{ij} = x_i^*$ for every maximizer $q_i$ of the UPM problem. Furthermore, for any provider $j$ such that $q_{ij} > 0$, $\frac{p_j}{c_{ij}} = \min_{k \in \mathcal{J}} \frac{p_k}{c_{ik}}$.*

However, not every user requests resources from a single provider. This leads to the following definition.

**Definition 6.17    (Preference set)**    For any price vector $p$, user $i$'s preference set $\mathcal{J}_i(p)$ includes each provider $j \in \mathcal{J}$ with $\frac{p_j}{c_{ij}} = \min_{k \in \mathcal{J}} \frac{p_k}{c_{ik}}$.

In light of Lemma 6.16, $\mathcal{J}_i$ is the set of providers from which user $i$ might request a strictly positive amount of resource. Users can be partitioned to *decided users* and *undecided users* based on the cardinality of their preference sets. The preference set of a decided user $i$ contains a singleton, and there is a unique vector $q_i$ that maximizes his payoff. By contrast, the preference set of an undecided user $i$ contains more than one provider, and any choice of $q_i \geq 0$ such that $x_i^* = \sum_{j \in \mathcal{J}_i}^{¥} q_{ij} c_{ij}$ maximizes his payoff.

However, we can construct a bipartite graph representation (BGR) of the undecided users' preference sets as follows. We represent undecided users by circles, and providers of undecided users as squares. We place an edge $(i, j)$ between a provider node $j$ and a user node $i$ if $j \in \mathcal{J}_i$. Then we can show that this BGR has no loops with probability 1. This means that we can use the BGR to uniquely determine (decode) the demands of undecided users.

Suppose that the optimal Lagrange multipliers $p^*$ from Section 4.3.2 are announced by providers as prices in this game. Since all users have complete network information, each of them can calculate all users' preference sets, and can construct the corresponding BGR. Undecided users can now uniquely determine their demand vectors by independently running the same BGR decoding algorithm. The demand $q^*$ found through BGR decoding will be unique. We can in fact show that this corresponds to the unique subgame perfect equilibrium of the provider competition game. For details, see [31].

**Equilibrium strategy of the providers in Stage I**

The optimal choice of prices for the providers depends on how the users' demand changes with respect to the price, which further depends on the users' utility functions. The quantity that indicates how a user's demand changes with respect to the price is the *coefficient of relative risk aversion* [18] of the utility function $u_i$, i.e., $k_{RRA}^i = -x u_i''(x)/u_i'(x)$. We focus on a class of utility functions characterized in Assumption 6.18.

**Assumption 6.18**    For each user $i \in \mathcal{I}$, the coefficient of relative risk aversion of its utility function is less than 1.

Assumption 6.18 is satisfied by some commonly used utility functions, such as $\log(1 + x)$ and the $\alpha-$fair utility functions $x^{1-\alpha}/(1 - \alpha)$ for $\alpha \in (0, 1)$. Under Assumption 6.18, we can show that a monopolistic provider will sell all of its resource $Q_j$ to maximize its revenue. The intuition is that Assumption 6.18 ensures that user demands are elastic. This encourages a provider to charge the lowest price (and hence sell all the resources) to maximize his revenue. We show that this is true even if multiple providers are competing with each other.

**Theorem 6.19**  *Under Assumptions 4.12, 4.13, and 4.14 in Section 4.3.1 and Assumption 6.18 above, the unique socially optimal demand vector $q^*$ and the associated Lagrangian multiplier vector $p^*$ of the SWO problem constitute the unique sub-game perfect equilibrium of the provider competition game.*

It is interesting to see that the competition of providers does not reduce social efficiency. This is not a simple consequence of the strict concavity of the users' utility functions; it is also related to the elasticity of users' demands.

Figure 6.8 summarizes the three sets of concepts discussed in Sections 4.3.2, 4.3.3, and 6.3.1.

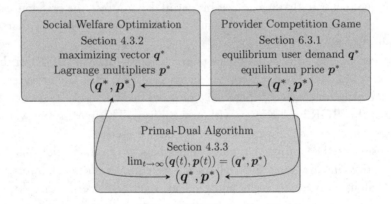

**Figure 6.8:** Relationship between different concepts for wireless service provider competition.

# 6.4 APPLICATION II: COMPETITION WITH SPECTRUM LEASING

Wireless spectrum is often considered as a scarce resource, and thus has been tightly controlled by the government through static license-based allocations. However, several recent field measurements show that many spectrum bands are often under utilized even in densely populated urban areas. To achieve more efficient spectrum utilization, secondary users may be allowed to share the spectrum with the licensed primary users. Various dynamic spectrum access mechanisms have been proposed along this direction. One of the proposed mechanisms is dynamic spectrum leasing, where a spectrum owner dynamically transfers and trades the usage right of a temporarily unused part of its licensed

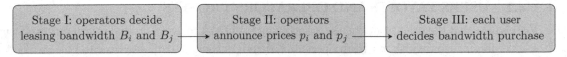

| Stage I: operators decide leasing bandwidth $B_i$ and $B_j$ | → | Stage II: operators announce prices $p_i$ and $p_j$ | → | Stage III: each user decides bandwidth purchase |

**Figure 6.9:** Three-stage dynamic game: the duopoly's leasing and pricing, and the users' resource allocation.

spectrum to secondary network operators or users in exchange for monetary compensation. In this application, we study the competition of two secondary operators under the dynamic spectrum leasing paradigm.

### 6.4.1 NETWORK MODEL

We consider two operators ($i, j \in \{1, 2\}$ and $i \neq j$) and a set $\mathcal{K} = \{1, \ldots, K\}$ of users in an ad hoc network. The operators obtain wireless spectrum from different spectrum owners with different leasing costs, and compete to serve the same set $\mathcal{K}$ of users. Each user has a transmitter-receiver pair. We assume that users are equipped with software defined radios (SDRs) and can transmit in a wide range of frequencies as instructed by the operators. A user may switch among different operators' services (e.g., WiMAX, 3G) depending on operators' prices. It is important to study the competition among multiple operators as operators are normally not cooperative.

The interactions between the two operators and the users can be modeled as a *three-stage dynamic game*, as shown in Fig. 6.9. Operators $i$ and $j$ first simultaneously determine their leasing bandwidths in Stage I, and then simultaneously announce the prices to the users in Stage II. Finally, each user chooses to purchase bandwidth from *only one operator* to maximize its payoff in Stage III.

Here are several key notations for our problem:

- *Leasing decisions $B_i$ and $B_j$*: leasing bandwidths of operators $i$ and $j$ in Stage I, respectively.

- *Costs $C_i$ and $C_j$*: the fixed positive leasing costs per unit bandwidth for operators $i$ and $j$, respectively. These costs are determined by the negotiation between the operators and their own spectrum suppliers.

- *Pricing decisions $p_i$ and $p_j$*: prices per unit bandwidth charged by operators $i$ and $j$ to the users in Stage II, respectively.

- *A user $k$'s demand $w_{ki}$ or $w_{kj}$*: the bandwidth demand of a user $k \in \mathcal{K}$ from operator $i$ or $j$. A user can only purchase bandwidth from one operator.

### 6.4.2   USERS' PAYOFFS AND OPERATORS' PROFITS

We assume that the users share the spectrum using OFDMA to avoid mutual interferences. If a user $k \in \mathcal{K}$ obtains bandwidth $w_{ki}$ from operator $i$, then it achieves a data rate (in nats) of

$$r_k(w_{ki}) = w_{ki} \ln \left( \frac{P_k^{\max} h_k}{n_0 w_{ki}} \right), \tag{6.4}$$

where $P_k^{\max}$ is user $k$'s maximum transmission power, $n_0$ is the noise power density, $h_k$ is the channel gain between user $k$'s transmitter and receiver. The channel gain $h_k$ is independent of the operator, as the operator only sells bandwidth and does not provide a physical infrastructure. We also assume that each user experiences a flat fading over the entire spectrum, such as in the current 802.11d/e standard where the channels are formed through proper interleaving. Here we assume that user $k$ spreads its power $P_k^{\max}$ across the entire allocated bandwidth $w_{ki}$. Furthermore, we focus on the high SNR regime where $\text{SNR} \gg 1$, such that Shannon capacity $\ln(1 + \text{SNR})$ can be approximated by $\ln(\text{SNR})$. To simplify later discussions, we let

$$g_k = \frac{P_k^{\max} h_k}{n_0},$$

thus $g_k/w_{ki}$ is the user $k$'s SNR.

If a user $k$ purchases bandwidth $w_{ki}$ from operator $i$, it receives a *payoff* of

$$u_k(p_i, w_{ki}) = w_{ki} \ln \left( \frac{g_k}{w_{ki}} \right) - p_i w_{ki}, \tag{6.5}$$

which is the difference between the data rate and the payment. The payment is proportional to price $p_i$ announced by operator $i$.

For an operator $i$, its profit is the difference between the revenue and the total cost, i.e.,

$$\pi_i(B_i, B_j, p_i, p_j) = p_i Q_i(B_i, B_j, p_i, p_j) - B_i C_i, \tag{6.6}$$

where $Q_i(B_i, B_j, p_i, p_j)$ and $Q_j(B_i, B_j, p_i, p_j)$ are realized demands of operators $i$ and $j$. The concept of realized demand will be defined later in Definition 6.23.

### 6.4.3   ANALYSIS OF THE THREE-STAGE GAME

We will use backward induction to compute the subgame perfect equilibrium (SPE). We will start with Stage III and analyze the users' behaviors given the operators' investment and pricing decisions. Then we look at Stage II and analyze how operators make the pricing decisions taking the users' demands in Stage III into consideration. Finally, we look at the operators' leasing decisions in Stage I using the results in Stages II and III.

In the following analysis, we only focus on pure strategy SPE and rule out mixed SPE in the multi-stage game. We say a *conditionally* SPE [47] is an SPE with pure strategies only, where the network's pure strategies constitute a Nash equilibrium in every subgame. In the following analysis, we derive the conditionally SPE, which is also referred to as an equilibrium for simplicity.

**Spectrum Allocation in Stage III**

In Stage III, each user needs to decide how much spectrum to purchase from which operator, based on the prices $p_i$ and $p_j$ announced by the operators in Stage II.

If a user $k \in \mathcal{K}$ obtains bandwidth $w_{ki}$ from operator $i$, then its payoff $u_k(p_i, w_{ki})$ is given in (6.5). Since this payoff is concave in $w_{ki}$, the unique *demand* that maximizes the payoff is

$$w_{ki}^*(p_i) = \arg\max_{w_{ki} \geq 0} u_k(p_i, w_{ki}) = g_k \exp(-(1 + p_i)). \tag{6.7}$$

Demand $w_{ki}^*(p_i)$ is always positive, linear in $g_k$, and decreasing in price $p_i$. Since $g_k$ is linear in channel gain $h_k$ and transmission power $P_k^{\max}$, we have that a user with a better channel condition or a larger transmission power has a larger demand.

Next we explain how each user decides which operator to purchase from. The following definitions help the discussions.

**Definition 6.20  (Preferred User Set)**  The Preferred User Set $\mathcal{K}_i^P$ includes the users who prefer to purchase from operator $i$.

**Definition 6.21  (Preferred Demand)**  The Preferred Demand $D_i$ is the total demand from users in the preferred user set $\mathcal{K}_i^P$, i.e.,

$$D_i(p_i, p_j) = \sum_{k \in \mathcal{K}_i^P(p_i, p_j)} g_k \exp(-(1 + p_i)). \tag{6.8}$$

The notations in (6.8) imply that both set $\mathcal{K}_i^P$ and demand $D_i$ only depend on prices $(p_i, p_j)$ in Stage II and are independent of operators' leasing decisions $(B_i, B_j)$ in Stage I. Such dependence can be discussed in two cases:

1. *Different Prices $(p_i < p_j)$*: every user $k \in \mathcal{K}$ *prefers* to purchase from operator $i$ since

$$u_k(p_i, w_{ki}^*(p_i)) > u_k(p_j, w_{kj}^*(p_j)).$$

   We have $\mathcal{K}_i^P = \mathcal{K}$ and $\mathcal{K}_j^P = \emptyset$, and

$$D_i(p_i, p_j) = G \exp(-(1 + p_i)) \text{ and } D_j(p_i, p_j) = 0,$$

   where $G = \sum_{k \in \mathcal{K}} g_k$ represents the aggregate wireless characteristics of the users.

2. *Same Prices $(p_i = p_j = p)$*: every user $k \in \mathcal{K}$ is indifferent between the operators and randomly chooses one with equal probability. In this case,

$$D_i(p, p) = D_j(p, p) = G \exp(-(1 + p))/2.$$

Now let us discuss how much demand an operator can actually satisfy, which depends on the bandwidth investment decisions $(B_i, B_j)$ in Stage I.

It is useful to define the following terms.

**Definition 6.22   (Realized User Set)**   The Realized User Set $\mathcal{K}_i^R$ includes the users whose demands are satisfied by operator $i$.

**Definition 6.23   (Realized Demand)**   The Realized Demand $Q_i$ is the total demand of users in the Realized User Set $\mathcal{K}_i^R$, i.e.,

$$Q_i\left(B_i, B_j, p_i, p_j\right) = \sum_{k \in \mathcal{K}_i^R(B_i, B_j, p_i, p_j)} g_k \exp(-(1 + p_i)) \, .$$

Notice that both $\mathcal{K}_i^R$ and $Q_i$ depend on prices $(p_i, p_j)$ in Stage II and leasing decisions $(B_i, B_j)$ in Stage I. Calculating the Realized Demands also requires considering two different pricing cases.

1. *Different prices $(p_i < p_j)$:* The Preferred Demands are $D_i(p_i, p_j) = G \exp(-(1 + p_i))$ and $D_j(p_i, p_j) = 0$.

   - *If Operator i has enough resource $\left(i.e., B_i \geq D_i\left(p_i, p_j\right)\right)$:* all Preferred Demand will be satisfied by operator $i$. The Realized Demands are

$$\begin{aligned} Q_i &= \min(B_i, D_i(p_i, p_j)) = G \exp(-(1 + p_i)), \\ Q_j &= 0. \end{aligned}$$

   - *If Operator i has limited resource $\left(i.e., B_i < D_i\left(p_i, p_j\right)\right)$:* since operator $i$ cannot satisfy the Preferred Demand, some demand will be satisfied by operator $j$ if it has enough resource. Since the realized demand $Q_i(B_i, B_j, p_i, p_j) = B_i = \sum_{k \in \mathcal{K}_i^R} g_k \exp(-(1 + p_i))$, then $\sum_{k \in \mathcal{K}_i^R} g_k = B_i \exp(1 + p_i)$.[3] The remaining users want to purchase bandwidth from operator $j$ with a total demand of $(G - B_i \exp(1 + p_i)) \exp(-(1 + p_j))$. Thus, the Realized Demands are

$$\begin{aligned} Q_i &= \min(B_i, D_i(p_i, p_j)) = B_i, \\ Q_j &= \min\left(B_j, \frac{G - B_i \exp(1 + p_i)}{\exp(1 + p_j)}\right). \end{aligned}$$

[3]Here we consider a large number of users and each user is non-atomic (infinitesimal). Thus, an individual user's demand is infinitesimal to an operator's supply and we can claim equality holds for $Q_i = B_i$.

2. *Same prices* $(p_i = p_j = p)$: both operators will attract the same Preferred Demand $G \exp(-(1+p))/2$. The Realized Demands are

$$
\begin{aligned}
Q_i &= \min\left(B_i, \frac{G}{2\exp(1+p)} + \max\left(\frac{G}{2\exp(1+p)} - B_j, 0\right)\right), \\
Q_j &= \min\left(B_j, \frac{G}{2\exp(1+p)} + \max\left(\frac{G}{2\exp(1+p)} - B_i, 0\right)\right).
\end{aligned}
$$

**Operators' Pricing Competition in Stage II**

In Stage II, the two operators simultaneously determine their prices $(p_i, p_j)$ considering the users' preferred demands in Stage III, given the investment decisions $(B_i, B_j)$ in Stage I.

An operator $i$'s profit is defined earlier in (6.6). Since the payment $B_i C_i$ is fixed at this stage, operator $i$'s profit maximization problem is equivalent of maximization of its revenue $p_i Q_i$. Note that users' total demand $Q_i$ to operator $i$ depends on the received power of each user (product of its transmission power and channel gain). We assume that an operator $i$ knows users' transmission powers and channel conditions. This can be achieved in a similar way as it is in today's cellular networks, where users need to register with the operator when they enter the network and frequently feedback the channel conditions. Thus, we assume that an operator knows the user population and user demand.

**Game 6.24  Pricing Game**   The competition between the two operators in Stage II can be modeled as the following game:

- Players: two operators $i$ and $j$.

- Strategy space: operator $i$ can choose price $p_i$ from the feasible set $\mathcal{P}_i = [0, \infty)$. Similarly, for operator $j$.

- Payoff function: operator $i$ wants to maximize the revenue $p_i Q_i(B_i, B_j, p_i, p_j)$. Similarly, for operator $j$.

At an equilibrium of the pricing game, $(p_i^*, p_j^*)$, each operator maximizes its payoff assuming that the other operator chooses the equilibrium price, i.e.,

$$
p_i^* = \arg\max_{p_i \in \mathcal{P}_i} p_i Q_i(B_i, B_j, p_i, p_j^*), \quad i = 1, 2, i \neq j .
$$

In other words, no operator wants to unilaterally change its pricing decision at an equilibrium.

Next we will investigate the existence and uniqueness of the pricing equilibrium. First, we show that it is sufficient to only consider symmetric pricing equilibrium for Game 6.24.

**Proposition 6.25**   *Assume both operators lease positive bandwidths in Stage I, i.e., $\min(B_i, B_j) > 0$. If a pricing equilibrium exists, it must be symmetric, i.e., $p_i^* = p_j^*$.*

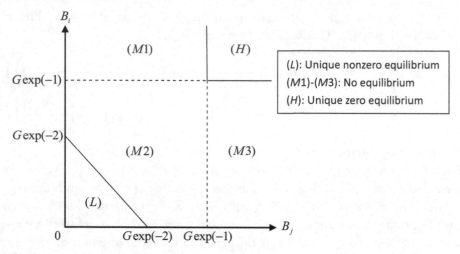

**Figure 6.10:** Pricing equilibrium types in different $(B_i, B_j)$.

The intuition is that no operator will announce a price higher than its competitor to avoid losing its Preferred Demand. This property significantly simplifies the search for all possible equilibria.

Next we show that the symmetric pricing equilibrium is a function of $(B_i, B_j)$ as shown in Fig. 6.10.

**Theorem 6.26**   *The equilibria of the pricing game are as follows.*

- *Low Investment Regime ($B_i + B_j \leq G \exp(-2)$ as in region (L) of Fig. 6.10): there exists a unique nonzero pricing equilibrium*

$$p_i^*(B_i, B_j) = p_j^*(B_i, B_j) = \ln\left(\frac{G}{B_i + B_j}\right) - 1. \tag{6.9}$$

*The operators' profits in Stage II are*

$$\pi_{II,i}(B_i, B_j) = B_i\left(\ln\left(\frac{G}{B_i + B_j}\right) - 1 - C_i\right), \tag{6.10}$$

$$\pi_{II,j}(B_i, B_j) = B_j\left(\ln\left(\frac{G}{B_i + B_j}\right) - 1 - C_j\right). \tag{6.11}$$

*denotes the low investment regime.*

- *Medium Investment Regime ($B_i + B_j > G \exp(-2)$ and $\min(B_i, B_j) < G \exp(-1)$ as in regions (M1)-(M3) of Fig. 6.10): there is no pricing equilibrium.*

- *High Investment Regime (*$\min(B_i, B_j) \geq G \exp(-1)$ *as in region (H) of Fig. 6.10): there exists a unique zero pricing equilibrium*

$$p_i^*(B_i, B_j) = p_j^*(B_i, B_j) = 0,$$

*and the operators' profits are negative for any positive values of $B_i$ and $B_j$.*

Intuitively, higher investments in Stage I will lead to lower equilibrium prices in Stage II. Theorem 6.26 shows that the only interesting case is the low investment regime where both operators' total investment is no larger than $G \exp(-2)$, in which case there exists a unique positive symmetric pricing equilibrium. Notice that same prices at equilibrium do not imply same profits, as the operators can have different costs ($C_i$ and $C_j$) and thus can make different investment decisions ($B_i$ and $B_j$) as shown next.

### Operators' Leasing Strategies in Stage I

In Stage I, the operators need to decide the leasing amounts ($B_i, B_j$) to maximize their profits. Based on Theorem 6.26, we only need to consider the case where the total bandwidth of both the operators is no larger than $G \exp(-2)$. We emphasize that the analysis of Stage I is not limited to the case of low investment regime; we actually also consider the medium investment regime and the high investment regime. The key observation is that an (conditionally) SPE will not include any investment decisions ($B_i, B_j$) in the medium investment regime, as it will not lead to a pricing equilibrium in Stage II. Moreover, any investment decisions in the high investment regime lead to zero operator revenues and are strictly dominated by any decisions in low investment regime. After the above analysis, the operators only need to consider possible equilibria in the low investment regime in Stage I.

**Game 6.27 Investment Game**   The competition between the two operators in Stage I can be modeled as the following game:

- Players: two operators $i$ and $j$.

- Strategy space: the operators will choose ($B_i, B_j$) from the set $\mathcal{B} = \{(B_i, B_j) : B_i + B_j \leq G \exp(-2)\}$. Notice that the strategy space is coupled across the operators, but the operators do not cooperate with each other.

- Payoff function. the operators want to maximize their profits in (6.10) and (6.11), respectively.

At an equilibrium of the investment game, ($B_i^*, B_j^*$), each operator has maximized its payoff assuming that the other operator chooses the equilibrium investment, i.e.,

$$B_i^* = \arg \max_{0 \leq B_i \leq \frac{G}{\exp(2)} - B_j^*} \pi_{II,i}(B_i, B_j^*), \quad i = 1, 2, i \neq j.$$

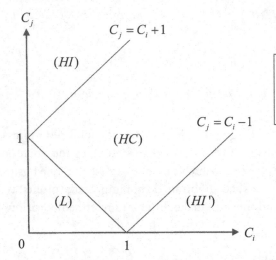

**Figure 6.11:** Leasing equilibrium types in different $(C_i, C_j)$.

To calculate the investment equilibria of Game 6.27, we can first calculate operator $i$'s best response given operator $j$'s (not necessarily equilibrium) investment decision, i.e.,

$$B_i^*(B_j) = \arg \max_{0 \le B_i \le \frac{G}{\exp(2)} - B_j} \pi_{II,i}(B_i, B_j), \quad i = 1, 2, i \ne j.$$

By looking at operator $i$'s profit in (6.10), we can see that a larger investment decision $B_i$ will lead to a smaller price. The best choice of $B_i$ will achieve the best tradeoff between a large bandwidth and a small price.

After obtaining best investment responses of duopoly, we can then calculate the investment equilibria, given different costs $C_i$ and $C_j$.

**Theorem 6.28** *The duopoly investment (leasing) equilibria in Stage I are summarized as follows.*

- *Low Costs Regime ($0 < C_i + C_j < 1$, as region (L) in Fig. 6.11): there exist infinitely many investment equilibria characterized by*

$$B_i^* = \rho G \exp(-2), \quad B_j^* = (1 - \rho)G \exp(-2), \tag{6.12}$$

*where $\rho$ can be any value that satisfies*

$$C_j \le \rho \le 1 - C_i. \tag{6.13}$$

*The operators' profits are*

$$\pi_{I,i} = B_i^*(1 - C_i), \quad \pi_{I,j} = B_j^*(1 - C_j).$$

- *High Comparable Costs Regime ($C_i + C_j \geq 1$ and $|C_j - C_i| \leq 1$, as region (HC) in Fig. 6.11): there exists a unique investment equilibrium*

$$B_i^* = \frac{(1 + C_j - C_i)G}{2} \exp\left(-\frac{C_i + C_j + 3}{2}\right),$$ (6.14)

$$B_j^* = \frac{(1 + C_i - C_j)G}{2} \exp\left(-\frac{C_i + C_j + 3}{2}\right).$$ (6.15)

*The operators' profits are*

$$\pi_{I,i} = \left(\frac{1 + C_j - C_i}{2}\right)^2 G \exp\left(-\left(\frac{C_i + C_j + 3}{2}\right)\right),$$

$$\pi_{I,j} = \left(\frac{1 + C_i - C_j}{2}\right)^2 G \exp\left(-\left(\frac{C_i + C_j + 3}{2}\right)\right).$$

- *High Incomparable Costs Regime ($C_j > 1 + C_i$ or $C_i > 1 + C_j$, as regions (HI) and (HI') in Fig. 6.11): For the case of $C_j > 1 + C_i$, there exists a unique investment equilibrium with*

$$B_i^* = G \exp(-(2 + C_i)), \quad B_j^* = 0,$$

*i.e., operator i acts as the monopolist and operator j is out of the market. The operators' profits are*

$$\pi_{I,i} = G \exp(-(2 + C_i)), \quad \pi_{I,j} = 0.$$

*The case of $C_i > 1 + C_j$ can be analyzed similarly.*

## 6.5 CHAPTER SUMMARY

This chapter discussed how multiple players compete with each other in a market, where pricing is one of the major decisions to make.

To understand the competition, we first introduced the basis of game theory. Game theory describes how multiple strategic players make their decisions to maximize their own payoffs, by taking the other players' decisions into consideration. We introduced the basics of noncooperative static and dynamic games with complete information. We first introduced the strategic form game, which is often used to model the simultaneous decisions of all players. We defined several important concepts including strictly dominated strategies, best response correspondence, and the Nash equilibrium. We further differentiated between pure strategy Nash equilibria and mixed strategy Nash equilibria, and showed several classical existence results. Then we moved on to introduce the extensive form game, where players make sequential decisions (also called a dynamic game). In such a game, the game history becomes very important, and the strategy is no longer a single action but a contingency plan based on the game history. We further introduced the concept of subgame perfect equilibrium, which is a generalization of the Nash equilibrium in a dynamic game.

With the knowledge of game theory, we are able to understand the oligopoly models, which characterize the competition between multiple firms in the same market. We introduced three types of oligopoly models: the Cournot model where firms compete based on quantity, the Bertrand model where firms compete based on pricing, and the Hotelling model that captures the impact of locations on the competition. Although we have used two firms (duopoly) as examples when introducing these three models, the results can be easily generalized to the case of more than two firms (oligopoly).

We illustrated the theory using two examples. The first one revisits the wireless service provider competition in Section 4.3. Instead of looking at the social optimal pricing as in Section 4.3, here we study how multiple providers will set their prices to maximize their revenues, by considering the users' locations and other providers' prices into consideration. We modeled the interactions by a multi-leader-follower game. Perhaps the most surprising result is that when the utility function satisfies the proper conditions, we can show that the unique subgame perfect equilibrium of the game is exactly the same as the unique global optimal solution of the social welfare optimization problem. This result holds regardless of the number of providers in the network, and thus is quite general and encouraging in practice. The second application considers a duopoly between two secondary operators, who will decide their capacity investments through spectrum leasing and market competition through spectrum pricing. We modeled the interactions between the operators and users as a three-stage multi-leader-follower game, and derived the conditionally SPE with pure strategies in each stage of the game. It turns out that when the leasing costs are low for both providers, then they engage in severe market competition and there are infinitely many equilibria. When the leasing costs are higher, the market will have a unique equilibrium, either two operators sharing the market or only the lower cost operator dominates the market. For more details especially mathematical proofs related to the two applications, please see [31, 48].

## 6.6  EXERCISES

1. *Congestion Game.* Consider the following communication congestion game. Two mobile users transmit on the same channel, each deciding whether or not to transmit its data at a particular time slot. Each user will incur a cost $c$ from each transmission, which mainly includes the power cost, channel access fee, etc. Each user can achieve a revenue $R > c$ from each successful transmission, and a zero revenue if a collision occurs (i.e., if both users transmit at the same time). The above congestion game can be represented by the following payoff matrix, where rows denote the actions of user 1, and columns denote the actions of user 2.

|  | Transmit | Not Transmit |
|---|---|---|
| Transmit | $(-c, -c)$ | $(R - c, 0)$ |
| Not Transmit | $(0, R - c)$ | $(0, 0)$ |

- Determine whether the game has pure strategy equilibria or not. If it does, find the pure strategy Nash equilibria.

- Find the set of all mixed-strategy Nash equilibria.

2. *Power Competition Game.* Consider a wireless network with $n$ mobile users. Each user represents a dedicated pair of transmitter and receiver. All users transmit on the same channel simultaneously through CDMA, and hence cause mutual interferences. Each user determines its transmission power to maximize the received signal to interference plus noise ratio (SINR) (or equivalently the channel capacity). Let $p_i$ denote the transmission power of user $i$, and $G_{ij}$ denote the channel gain between the transmitter of user $i$ and the receiver of user $j$. The utility of user $i$ is defined as the achieved capacity, i.e.,

$$f_i(p_1, ..., p_n) = \log\left(1 + \frac{G_{ii} \cdot p_i}{\sigma^2 + \sum_{j \neq i} G_{ji} \cdot p_j}\right),$$

where $G_{ii} \cdot p_i$ is the received signal power of user $i$, $\sum_{j \neq i} G_{ji} \cdot p_j$ is the total interference power (on user $i$) from all other users, and $\sigma^2$ is the noise power. The social welfare as the aggregate utility of all users, i.e., $\sum_{i=1}^{n} f_i(p_1, ..., p_n)$.

- Derive the social optimality, i.e., the power vector $(p_1, ..., p_n)$ that maximizes the aggregate SINR of all users.

- Show that there is a unique pure-strategy Nash equilibrium for this power competition game, in which every user transmits on its maximum power.

- Show that the social optimal solution is different from the Nash equilibrium.

3. *Price-based Power Competition Game.* We extend the above power competition game by introducing a regulator. The regulator charges every user $i$ a unit price $\pi_i$ for every unit of transmission power. The utility of user $i$ is defined as the difference between the revenue from its achieved capacity and the payment to the regulator, i.e.,

$$f_i(p_1, ..., p_n) = \log\left(1 + \frac{G_{ii} \cdot p_i}{\sigma^2 + \sum_{j \neq i} G_{ji} \cdot p_j}\right) - p_i \cdot \pi_i.$$

Derive the Nash equilibrium for this power competition game.

CHAPTER 7

# Network Externalities

In the previous chapters, we assumed that a decision made by a consumer or producer has no external effects on the other consumers or producers who are not directly involved. In practice, however, there are many situations where external or third-party effects are important. In these situations, the third parties' actions lead to either benefits or costs to the players who are not involved directly. In economics, these external effects are termed as *externalities* [49, 50].

In this chapter, we will first introduce the theory of externality following [49, 50]. Then we present two applications. In the first application, users generate negative externalities to each other due to interferences. The key idea to resolve this issue is to internalize the externality through Pigovian tax. In the second application, a cellular operator decides when to upgrade his service from 3G to 4G considering upgrading cost, user switching cost, and the impact of revenue due to network effect.

## 7.1 THEORY: NETWORK EXTERNALITIES

In this section, we cover the basic concepts of externalities, and study some classic externalities. We will see how externalities can be a source of market or network inefficiency, and study some approaches to combat such inefficiencies.

### 7.1.1 WHAT IS EXTERNALITY?

Simply speaking, externalities are the benefits or costs that are imposed by the actions of one player on a third party not directly involved [49, 50]. The office mates who breathe the second-hand smoke, the wireless users who experience the interferences from the nearby transmitters, the shoppers who enjoy the department store Christmas displays—these are all good examples of consumers experiencing the costs or benefits imposed by other consumers. Such costs and benefits are said to be *external* and are thus called *externalities*. External costs (like the smoke and interference) are called *negative* externalities, while external benefits (like the pleasure from enjoying the Christmas decorations) are called *positive* externalities.

In the context of a free market, we can define externality as "any indirect effect that either a production or a consumption activity has on the utility function or the consumption set of a consumer, or the production set of a producer" [50]. By "indirect," it means that the effect concerns a third party other than the one who exerts this activity or who is involved directly, and the effect is not transferred through prices. More simply, an externality is an economic side effect.

**Definition 7.1  Externality**    An externality is any side effect (benefit or cost) that is imposed by the actions of a player on a third party not directly involved.

As mentioned previously, externalities may be positive or negative. A negative externality is also called the external cost, and a positive externality is also called the external benefit. Examples of negative externalities include *pollution* (such as air pollution, water pollution, and noise pollution) and *interference* in wireless networks. In the examples of pollution, the producer or consumer finances the goods produced, but the third-party society must bear the cost of pollution that is introduced into the environment as a by-product. In the example of wireless interference, the mobile user transmits his own data for benefits, but other users (who are not intended receivers) will suffer performance degradation due to the interference caused by this user. An example of positive externalities is the *network effect*. With network effect, more users consuming goods or services makes a good or service more valuable. Network effect is an important theme in telecommunication networks and online social networks. The more people own telephones or access to a social network, the more valuable the telephone or the social network is to each user, since he can connect to more people by his own telephone or through the social network.[1]

Externalities can cause market failure if the price mechanism does not take into account the external costs and external benefits of production and consumption. The reason is that the producers or consumers are interested in maximizing their profits only. Therefore, they will only take into account the private costs and private benefits arising from their supply or demand of the product, but not the social costs and social benefits. As a consequence, the producers (or consumers) profit maximizing the level of supply (or demand) will deviate from the social optimum level. We show this in Figure 7.1.

The left subfigure in Figure 7.1 illustrates the production deviation induced by the negative externality (or external cost) of production. The social cost includes not only the producers' private costs, but also the external costs. Thus, the social marginal cost is larger than the private marginal cost, as shown in the figure. Accordingly, the social optimum level of supply is $Q^*$, which is smaller than the producers' private profit maximizing level of supply $Q_1$. Similarly, the right subfigure illustrates the consumption deviation induced by the positive externality (or external benefit) of consumption. The social benefit (or revenue) includes not only the consumers' private benefits, but also the external benefits. Thus, the social marginal revenue is larger than the private marginal revenue, as shown in the figure. Accordingly, the social optimum level of demand is $Q^*$, which is larger than the consumers' private profit maximizing level of demand $Q_1$.

Next we discuss the two types of network externalities in more detail.

---

[1]The expression "network effect [53, 54]" is applied most commonly to positive network externalities as in telecommunication networks and online social networks.

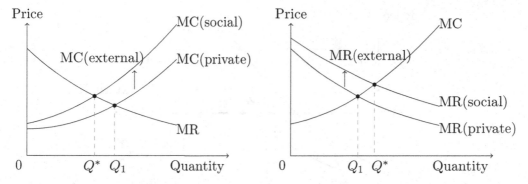

**Figure 7.1:** Market failures arising from (left) negative production externalities and (right) positive consumption externalities.

## 7.1.2   NEGATIVE EXTERNALITY

As shown in Figure 7.1, negative externalities may cause market failures such as over-production and over-consumption of the product. Many negative externalities are related to the environmental consequences of production and consumption.

**Example: Pollution**

In a pollution model, the producer benefits from his production activity, but a third party such as the society must bear the cost of pollution. Consumer's consumption activities can also cause pollution (like the smoker).

Consider a simple example of pollution. There are two firms: a chemical plant and a water company. The chemical plant produces chemical products and discharges wastewater into a river, which causes the water pollution in the river. Each unit chemical product is sold at a market equilibrium price of $10. That is, the marginal revenue of each unit of product is $10 for the chemical plant. The water company produces bottled water by drawing water from the river. The chemical plant's wastewater lowers the quality of the water, and therefore the water company must incur additional costs to purify the water before bottling it. Such an additional cost is the negative externality of the production activity of the chemical plant. Figure 7.2 shows the private marginal cost of the chemical plant, and the external marginal cost incurred by the water company. The social cost is the summation of both costs.

Since there is no incentive for the chemical plant to cover the external cost, it will choose the quantity of productions that maximizes its own profit, i.e., that equalizes its private marginal cost and marginal revenue (shown by the point $Q_1$ in the figure). However, this is not optimal from a social perspective. It is easy to derive that the social optimum quantity is $Q^*$, which equalizes the social marginal cost and the marginal revenue. Obviously, $Q^*$ is smaller than $Q_1$.

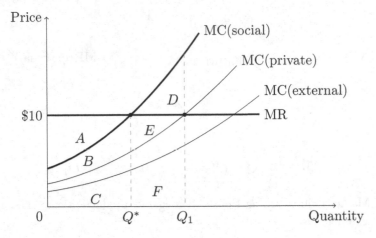

**Figure 7.2:** Marginal costs in the pollution model.

Now let's consider the chemical plant's and the water company's revenue (or cost) at different levels of quantity. As the chemical plant's profit maximizing quantity level $Q_1$, the chemical plant's total surplus is the sum of areas $A$, $B$, and $E$ (denoted by $A + B + E$). The water company's total surplus is $-(C + F)$ due to the external cost of water purifying. Notice that $B = C$ and $F = D + E$, since the social marginal cost is the sum of the private marginal cost of the chemical plant and the external cost incurred by the water company. Thus, the social surplus at quantity $Q_1$ is $A - D$. At the social optimal quantity level $Q^*$, the chemical plant's total surplus is $A + B$, and the water company's total surplus is $-C$. Thus, the social surplus at quantity $Q^*$ is $A$. This shows that with negative externalities, the individual profit maximizing decision may hurt the social surplus.

This example leaves us a key question: *is it possible to motivate the chemical plant to produce the social optimal quantity level in the absence of a centralized planner?* Most earlier economists argued that a decentralized competitive price system could reach the social optimum by either costlessly internalizing the externality by government or assessing taxes on the firm creating the negative externality. Due to space limits, we will discuss the second approach briefly as follows.

## Solution: Pigovian Tax

Pigovian tax, named after Arthur C. Pigou (1877–1959), is proposed to address the market failure caused by negative externalities. A Pigovian tax is a tax levied on a market activity that generates negative externalities [51]. A Pigovian tax equal to the negative externality can correct the market outcome back to efficiency.

To deal with over-production (or over-consumption similarly), Pigou recommended a tax placed on the offending producer (or consumer) for each unit of production (or consumption). If the government can accurately measure the external cost or social cost, the tax could equalize the marginal private cost and the marginal social cost, and therefore eliminate the market failure. In more

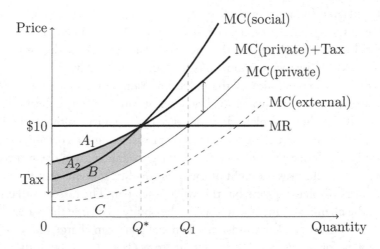

**Figure 7.3:** Illustration of the Pigovian tax.

specific terms, the producer would have to pay for the externality cost that it created. This would effectively reduce the quantity of the product produced, moving the economy back to an efficient equilibrium.

Figure 7.3 illustrates the effect of a Pigovian tax on the chemical plant's output. The tax shifts the private marginal cost curve up by the amount of the tax, and the shaded area $A_2 + B$ is the tax revenue. As shown in the figure, with the tax, the chemical plant has the incentive to reduce its output to the socially optimum quantity level $Q^*$. The chemical plant's surplus at quantity level $Q^*$ is $(A + B) - (A_2 + B) = A_1$, and the water company's total surplus is still $-C$.

Although this Pigovian tax works perfectly in theory, the practical implementation is very difficult due to a lack of complete information on the marginal social cost. Most of the criticism of the Pigovian tax relates to the determination of the marginal social cost and therefore the tax. In fact, Pigou himself also pointed out in [51] that the assumption that the government can determine the marginal social cost of a negative externality and convert that amount into a monetary value is a weakness of the Pigovian tax.

**Further Discussions: The Coase Theorem**

In the 1960s, Ronald Coase, a Nobel Prize winner, argued that the traditional analysis on externality was incomplete [52]. In terms of our example, Coase would say that the fundamental difficulty is not that the chemical plant creates an externality but that no one owns the quality of water. Coase would argue that the chemical plant will produce the socially optimal output, if (i) transaction costs are negligible, and (ii) one or the other party has clearly defined property rights in water quality. The *transaction cost* refers to the cost of negotiating, verifying, and enforcing contracts, and the *property right* is the exclusive authority to determine how a resource is used, whether that resource is owned by government, collective bodies, or by individuals. What was truly remarkable at the time was Coase's claim that the output produced by the chemical plant did not depend on which party possesses the property right. In other words, even if the chemical plant itself owns the property right of the water quality, it will produce the social optimal quantity level voluntarily!

We show the Coase's claim by the pollution example mentioned above. Suppose the water company owns the right of the water quality. Then it can costlessly collect external costs from the chemical plant, if the chemical plant degrades the water quality. In this case, the marginal cost for the chemical plant would exactly be the sum of its private marginal cost and the external cost. Thus, the chemical plant will select the social optimal output. Things are more surprising in the case that the chemical plant itself owns the right of the water quality. It seems that the chemical plant would over-produce to maximize its profit, but this will not happen. The reason is that with the right of the water quality, the chemical plant essentially has the ability to charge the water company for keeping the water high quality.[2] In other words, the chemical plant can charge the water company a certain amount of money for the decrease of its output from $Q_1$ to $Q^*$. Obviously, as long as the money is well designed, e.g., between $[E, D + E]$, both the chemical plant and water company has the incentive to accept such a deal.

Now let us go back to Coase's results. In the previous claims, Coase actually introduced an alternative approach to solve negative externalities, using the *property rights theory* [52]. Based on the property rights theory, Coase pointed in his famous article "The Problem of Social Cost [52]" that if trade in an externality is possible and there are no transaction costs, bargaining will lead to an efficient outcome regardless of the initial allocation of property rights. This is called the *Coase theorem*. Formally,

**Theorem 7.2 Coase Theorem** *As long as private property rights are well defined and transaction costs are negligible, exchange will eliminate divergence and lead to efficient use of resources or highest valued use of resources.*

Despite the perfection in theory, there are some criticisms about the practical application of the theorem, among which a key criticism is that the theorem is almost always inapplicable in economic reality, because real-world transaction costs are rarely low enough to allow for efficient bargaining.

### 7.1.3 POSITIVE EXTERNALITY

As shown in Figure 7.1, positive externalities may cause market failures such as under-production and under-consumption of the product. In computer networking, a typical positive externality is the network effect, where the higher usage of certain products makes them more valuable [53, 54].

---

[2]Such an ability is achieved by the chemical plant's threat of producing a large number of products (and accordingly degrading the water quality).

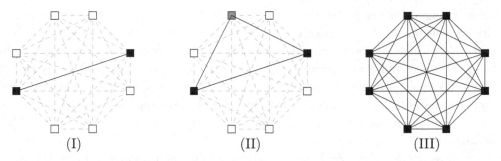

**Figure 7.4:** Illustration of Network Effect. (I) Each user gets one unit of benefit; (II) With the joining of a new user, each user gets two units of benefit; and (III) When all eight users join the network, each user gets seven units of benefit.

**Example: Network Effect**

A classic example of positive externalities is the *network effect* [53, 54]. In these situations, a product displays positive network effects when more usage of the product by any user increases the product's value for other users (and sometimes all users).

Network effect is one of the most important underlying economic concepts in industrial organization of IT industries, especially in telecommunication networks and online social networks. Network effects were first studied in the context of long-distance telephony in the early 1970s (one of the earliest papers on the topic is [55]). Today, they are widely recognized as a critical aspect of the industrial organization of IT industries, and are prevalent in a wide variety of sectors, including software, microprocessors, telecommunications, e-commerce, and electronic marketplaces. Empirical evidence of network effects has been found in product categories as diverse as spreadsheets, databases, networking equipment, and DVD players.

Consider a very simple example of network effects in telecommunication networks, where each user's benefit is simply defined as the number of users connected. The more people own telephones, the more valuable the telephone is to each owner. This creates a positive externality because a user may purchase a telephone without the intention of creating more value for other users, but does so in any case. The reason is that the purchasing activity of one user essentially increases the range that other users can connect to, and accordingly increase the value of other users' telephones. We show this in Figure 7.4, where the solid square denotes the user with a telephone, and the hollow square denotes the user without telephone. When there are only two users in the network, each user can only connect to one person (as shown in subfigure I). If a third person joins the network, each of the two earlier users benefit from the joining of the third user, since now each of them can connect to two persons (as shown in subfigure II). We can similarly see the increase of network value to every existing user as more users join the network (as shown in subfigure III).

To generalize the above discussions further, let us consider a network with $N$ users, where each user perceives a value that increases with $N$. If each user attaches the same value to the possibility of connecting to any one of the other $N - 1$ users, it may be considered that he perceives a network value proportional to $N - 1$. Then the total value of the network is proportional to $N(N - 1)$, or roughly $N^2$, which is known as the Metcalfe's Law [56]. A refined model was suggested by Briscoe et al. [57], where each user perceives a value of order $\log(N)$. In that model, a user ranks the other users in decreasing order of importance and assigns a value $1/k$ to the $k$-th user in that order, for a total value $1 + 1/2 + \cdots + 1/(N - 1) \approx \log(N)$. The resulting total network value is $N \log(N)$, which is appropriate for cellular networks shown by quantitative studies [57]. This will be useful in the discussions of Section 7.3.

Network effects become significant after a certain subscription percentage has been achieved, called *critical mass*. At the critical mass point, the value obtained from the good or service is greater than or equal to the price paid for the good or service. As the value of the good is determined by the user base, this implies that after a certain number of people have subscribed to the service or purchased the good, additional people will subscribe to the service or purchase the good due to the value exceeding the price.

Thus, a key business concern will be: *how to attract users prior to reaching critical mass*. One way is to rely on extrinsic motivation, such as a payment, a fee waiver, or a request for friends to sign up. A more natural strategy is to build a system that has enough value without network effects, at least to early adopters. Then, as the number of users increases, the system becomes even more valuable and is able to attract a wider user base. This issue is particularly important for online social networks, whose value greatly relies on the number of subscribed users in the network. This is also the reason why the QQ (an online social communicating tool) of Tencent company is so popular in China. Even though Tencent company sometimes provides poor service and charges a higher price (for some applications), many new people still prefer the QQ network over other alternatives, because its large number of subscribers (estimated to be 798.2 million by March 2013) make it more valuable than other social networks in China. Another example is Facebook which went public in 2012. The high valuation of Facebook is largely because of its enormous number of subscribers around the world (estimated to be 1.11 billion by March 2013).

**Different Types of Network Effect**

There are many ways to classify networks effects. One popular segmentation views network effects as being of four kinds as shown below [53, 54].

1. *Direct network effects*. The simplest network effects are direct: increases in usage lead to direct increases in value. The original example of telephone service is a good illustration of this kind. Another example is online social networks, where users directly benefit from the participation of other users.

2. *Indirect network effects*. Network effects may also be indirect, where increased usage of one product spawns the production of increasingly valuable complementary goods, and this in turn results in an increase in the value of the original product. Examples of complementary goods include software (such as an Office suite for operating systems) and DVDs (for DVD

players). This is why Windows and Linux might compete not just for users, but also for software developers.

3. *Two-sided network effects.* Network effects can also be two-sided, where the usage increase by one set of users increases the value of a complementary product to another distinct set of users, and vice versa. Hardware/software platforms, reader/writer software pairs, marketplaces, and matching services all display these kinds of network effects. In many cases, one may think of indirect network effects as a one-directional version of two-sided network effects.

4. *Local network effects.* The structure of an underlying social network affects who benefits from whom. For example, a good displays local network effects when each consumer is influenced directly by the decisions of only a typically small subset of other consumers, instead of being influenced by the increase of the total number of consumers. Instant messaging is an example of a product that displays local network effects.

## 7.2   APPLICATION I: DISTRIBUTED WIRELESS INTERFERENCE COMPENSATION

Interference mitigation is an important problem in wireless networks. A basic technique for this issue is to control the nodes' transmit powers. In an ad hoc wireless network, the power control is complicated by the lack of centralized infrastructure, which necessitates the use of distributed approaches. This application addresses the distributed power control for rate-adaptive users in a wireless network. We consider a CDMA network, where all users spread their power over a single frequency band. The transmission rate of each user depends on its received signal-to-interference plus noise ratio (SINR). Our objective is to coordinate the power levels of all users to optimize the overall performance, which is measured in terms of the total network utility. To achieve this, we propose a protocol in which the users exchange price signals that indicate the negative externality of the received interference.

Because we assume that the users cooperate, we ignore incentive issues, which may occur in networks with non-cooperative users. For example, in a non-cooperative scenario, a user may attempt to manipulate its announced interference prices to increase its own utility at the expense of the overall network utility. It can, of course, compromise the performance of the distributed algorithms presented here. We note that it may be possible to "hard wire" the power control algorithm into the handsets, making such a manipulation of price information difficult.

### 7.2.1   NETWORK MODEL

We consider a snapshot of an ad hoc network with a set $\mathcal{M} = \{1, ..., M\}$ of distinct node pairs. As shown in Fig. 7.5, each pair consists of one dedicated transmitter and one dedicated receiver. We use the terms "pair" and "user" interchangeably in the following. In this section, we assume that each user $i$ transmits through a CDMA scheme over the total bandwidth of $B$ Hz. Over the time period

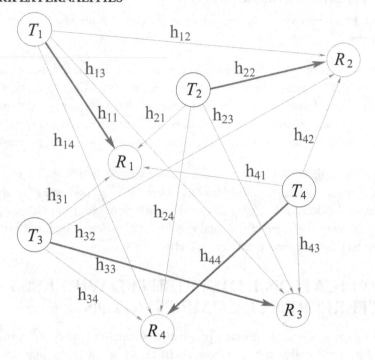

**Figure 7.5:** An example wireless network with four users (pairs of nodes) ($T_i$ and $R_i$ denote the transmitter and receiver of "user" $i$, respectively).

of interest, the channel gains of each pair are fixed. The channel gain between user $i$'s transmitter and user $j$'s receiver is denoted by $h_{ij}$. Note that in general $h_{ij} \neq h_{ji}$, since the latter represents the gain between user $j$'s transmitter and user $i$'s receiver.

Each user $i$'s quality of service is characterized by a utility function $u_i(\gamma_i)$, which is an increasing and strictly concave function of the received SINR,

$$\gamma_i(\boldsymbol{p}) = \frac{p_i h_{ii}}{n_0 + \frac{1}{B}\sum_{j \neq i} p_j h_{ji}}, \tag{7.1}$$

where $\boldsymbol{p} = (p_1, \cdots, p_M)$ is a vector of the users' transmission powers and $n_0$ is the background noise power. The users' utility functions are coupled due to mutual interference. An example utility function is a *logarithmic utility function* $u_i(\gamma_i) = \theta_i \log(\gamma_i)$, where $\theta_i$ is a user-dependent priority parameter.

The problem we consider is to specify $\boldsymbol{p}$ to maximize the utility summed over all users, where each user $i$ must also satisfy a transmission power constraint, $p_i \in \mathcal{P}_i = \left[ P_i^{\min}, P_i^{\max} \right]$, i.e.,

$$\max_{\{\boldsymbol{p}: p_i \in \mathcal{P}_i \, \forall i\}} \sum_{i=1}^{M} u_i(\gamma_i(\boldsymbol{p})). \tag{P1}$$

Note that a special case is $P_i^{\min} = 0$; i.e., the user may choose not to transmit.[3]

Although $u_i(\cdot)$ is concave, the objective in Problem P1 may not be concave in $p$. However, it is easy to verify that any local optimum, $p^* = (p_1^*, ..., p_M^*)$, of this problem will be regular (see p. 309 of [20]), and so must satisfy the Karush-Kuhn-Tucker (KKT) necessary conditions:

**Lemma 7.3  KKT conditions:**   *For any local maximum $p^*$ of Problem P1, there exist unique Lagrange multipliers $\lambda_{1,u}^*, ..., \lambda_{M,u}^*$ and $\lambda_{1,l}^*, ..., \lambda_{M,l}^*$ such that for all $i \in \mathcal{M}$,*

$$\frac{\partial u_i\left(\gamma_i\left(p^*\right)\right)}{\partial p_i} + \sum_{j \neq i} \frac{\partial u_j\left(\gamma_j\left(p^*\right)\right)}{\partial p_i} = \lambda_{i,u}^* - \lambda_{i,l}^*, \tag{7.2}$$

$$\lambda_{i,u}^*(p_i^* - P_i^{\max}) = 0, \ \lambda_{i,l}^*(P_i^{\min} - p_i^*) = 0, \ \lambda_{i,u}^*, \lambda_{i,l}^* \geq 0. \tag{7.3}$$

Let

$$\pi_j\left(p_j, p_{-j}\right) = -\frac{\partial u_j\left(\gamma_j\left(p_j, p_{-j}\right)\right)}{\partial I_j\left(p_{-j}\right)}, \tag{7.4}$$

where $I_j\left(p_{-j}\right) = \sum_{k \neq j} p_k h_{kj}$ is the total interference received by user $j$ (before bandwidth scaling). Here, $\pi_j\left(p_j, p_{-j}\right)$ is always nonnegative and represents user $j$'s marginal increase in utility per unit decrease in total interference. Using (7.4), condition (7.2) can be written as

$$\frac{\partial u_i\left(\gamma_i\left(p^*\right)\right)}{\partial p_i} - \sum_{j \neq i} \pi_j\left(p_j^*, p_{-j}^*\right) h_{ij} = \lambda_{i,u}^* - \lambda_{i,l}^*. \tag{7.5}$$

Viewing $\pi_j\left(= \pi_j\left(p_j, p_{-j}\right)\right)$ as a *price* charged to other users for generating interference to user $j$, condition (7.5) is a necessary and sufficient optimality condition for the problem in which each user $i$ specifies a power level $p_i \in \mathcal{P}_i$ to maximize the following surplus function

$$s_i\left(p_i; p_{-i}, \pi_{-i}\right) = u_i\left(\gamma_i\left(p_i, p_{-i}\right)\right) - p_i \sum_{j \neq i} \pi_j h_{ij}, \tag{7.6}$$

assuming fixed $p_{-i}$ and $\pi_{-i}$ (i.e., each user is a price taker and ignores any influence he may have on these prices). User $i$ therefore maximizes the difference between its utility minus its payment to the other users in the network due to the interference it generates. The payment is its transmit power times a weighted sum of other users' prices, with weights equal to the channel gains between user $i$'s transmitter and the other users' receivers. This pricing interpretation of the KKT conditions motivates the following asynchronous distributed pricing (ADP) algorithm.

---

[3]Occasionally, for technical reasons, we require $P_i^{\min} > 0$; in these cases, $P_i^{\min}$ can be chosen arbitrarily small so that this restriction has little effect. Note that for certain utilities, e.g., $\theta_i \log(\gamma_i)$, all assigned powers must be strictly positive, since as $p_i \to 0$, the utility approaches $-\infty$.

---

**Algorithm 4** The ADP Algorithm

---

(1) INITIALIZATION: For each user $i \in \mathcal{M}$ choose some power $p_i(0) \in \mathcal{P}_i$ and price $\pi_i(0) \geq 0$.

(2) POWER UPDATE: At each $t \in T_{i,p}$, user $i$ updates its power according to

$$p_i(t) = \mathcal{W}_i\left(p_{-i}(t^-), \pi_{-i}(t^-)\right).$$

(3) PRICE UPDATE: At each $t \in T_{i,\pi}$, user $i$ updates its price according to

$$\pi_i(t) = \mathcal{C}_i\left(\mathbf{p}(t^-)\right).$$

---

### 7.2.2   ASYNCHRONOUS DISTRIBUTED PRICING (ADP) ALGORITHM

In the ADP algorithm, each user announces a single price and all users set their transmission powers based on the received prices. Prices and powers are asynchronously updated. For $i \in \mathcal{M}$, let $T_{i,p}$ and $T_{i,\pi}$, be two unbounded sets of positive time instances at which user $i$ updates its power and price, respectively. User $i$ updates its power according to

$$\mathcal{W}_i(p_{-i}, \pi_{-i}) = \arg\max_{\hat{p}_i \in \mathcal{P}_i} \; s_i\left(\hat{p}_i; p_{-i}, \pi_{-i}\right),$$

which corresponds to maximizing the surplus in (7.6). Each user updates its price according to

$$\mathcal{C}_i(\mathbf{p}) = -\frac{\partial u_i\left(\gamma_i\left(\mathbf{p}\right)\right)}{\partial I_i\left(p_{-i}\right)},$$

which corresponds to (7.4). Using these update rules, the ADP algorithm is given in Algorithm 4. Note that in addition to being asynchronous across users, each user also need not update its power and price at the same time.[4]

In the ADP algorithm not only are the powers and prices generated in a distributed fashion, but also each user only needs to acquire limited information. To see this note that the power update function can be written as[5]

$$\mathcal{W}_i(p_{-i}, \pi_{-i}) = \left[\frac{p_i}{\gamma_i\left(\mathbf{p}\right)} g_i\left(\frac{p_i}{\gamma_i\left(\mathbf{p}\right)}\left(\sum_{j \neq i} \pi_j h_{ij}\right)\right)\right]_{P_i^{\min}}^{P_i^{\max}},$$

---

[4]Of course, simultaneous updates of powers and prices per user and synchronous updating across all the users are just special cases of Algorithm 4.

[5]Notation $[x]_a^b$ means $\max\{\min\{x, b\}, a\}$.

where $\frac{p_i}{\gamma_i(\boldsymbol{p})}$ is independent of $p_i$, and

$$
g_i(x) = \begin{cases}
\infty, & 0 \le x \le u_i'(\infty), \\
(u_i')^{-1}(x), & u_i'(\infty) < x < u_i'(0), \\
0, & u_i'(0) \le x.
\end{cases}
$$

Likewise, the price update can be written as

$$
\mathcal{C}_i(\boldsymbol{p}) = \frac{\partial u_i(\gamma_i(\boldsymbol{p}))}{\partial \gamma_i(\boldsymbol{p})} \frac{(\gamma_i(\boldsymbol{p}))^2}{B p_i h_{ii}}.
$$

From these expressions, it can be seen that to implement the updates, each user $i$ only needs to know: $(i)$ its own utility $u_i$, the current SINR $\gamma_i$ and channel gain $h_{ii}$, $(ii)$ the "adjacent" channel gains $h_{ij}$ for $j \in \mathcal{M}$ and $j \ne i$, and $(iii)$ the price profile $\boldsymbol{\pi}$. By assumption each user knows its own utility. The SINR $\gamma_i$ and channel gain $h_{ii}$ can be measured at the receiver and fed back to the transmitter. Measuring the adjacent channel gains $h_{ij}$ can be accomplished by having each receiver periodically broadcast a beacon; assuming reciprocity, the transmitters can then measure these channel gains. The adjacent channel gains account for only $1/M$ of the total channel gains in the network; each user does not need to know the other gains. The price information could also be periodically broadcast through this beacon. Since each user announces only a single price, the number of prices scales linearly with the size of the network. Also, numerical results show that there is little effect on performance if users only convey their prices to "nearby" transmitters, i.e., those generating the strongest interference [58].

Denote the set of fixed points of the ADP algorithm by

$$
\mathcal{F}^{ADP} \equiv \{(\boldsymbol{p}, \boldsymbol{\pi}) \mid (\boldsymbol{p}, \boldsymbol{\pi}) = (\mathcal{W}(\boldsymbol{p}, \boldsymbol{\pi}), \mathcal{C}(\boldsymbol{p}))\}, \tag{7.7}
$$

where $\mathcal{W}(\boldsymbol{p}, \boldsymbol{\pi}) = (\mathcal{W}_k(p_{-k}, \pi_{-k}))_{k=1}^{M}$ and $\mathcal{C}(\boldsymbol{p}) = (\mathcal{C}_k(\boldsymbol{p}))_{k=1}^{M}$. Using the strict concavity of $u_i(\gamma_i)$ in $\gamma_i$, the following result can be easily shown.

**Lemma 7.4** *A power profile $\boldsymbol{p}^*$ satisfies the KKT conditions of Problem P1 (for some choice of Lagrange multipliers) if and only if $(\boldsymbol{p}^*, \mathcal{C}(\boldsymbol{p}^*)) \in \mathcal{F}^{ADP}$.*

If there is only one solution to the KKT conditions, then it must be the global maximum and the ADP algorithm would reach that point if it converges. In general, $\mathcal{F}^{ADP}$ may contain multiple points including local optima or saddle points.

## 7.2.3   CONVERGENCE ANALYSIS OF ADP ALGORITHM

We next characterize the convergence of the ADP algorithm by viewing it in a game theoretic context. We can consider a game where each player $i$'s strategy includes specifying both a power $p_i$ and a price $\pi_i$ to maximize a payoff equal to the surplus in (7.6). However, since there is no penalty for user $i$ announcing a high price, it can be shown that each user's best response is to choose a large enough price to force all other users to transmit at $P_i^{\min}$. This is certainly not a desirable outcome and suggests that the prices should be determined externally by another procedure. Instead, we consider the following *Fictitious Power-Price (FPP) control game*,

$$G_{FPP} = [\mathcal{FW} \cup \mathcal{FC}, \left\{\mathcal{P}_i^{\mathcal{FW}}, \mathcal{P}_i^{\mathcal{FC}}\right\}, \left\{s_i^{\mathcal{FW}}, s_i^{\mathcal{FC}}\right\}],$$

where the players are from the union of the sets $\mathcal{FW}$ and $\mathcal{FC}$, which are both copies of $\mathcal{M}$. $\mathcal{FW}$ is a *fictitious power player set*; each player $i \in \mathcal{FW}$ chooses a power $p_i$ from the strategy set $\mathcal{P}_i^{\mathcal{FW}} = \mathcal{P}_i$ and receives payoff

$$s_i^{\mathcal{FW}}(p_i; p_{-i}, \pi_{-i}) = u_i(\gamma_i(\boldsymbol{p})) - \sum_{j \neq i} \pi_j h_{ij} p_i. \tag{7.8}$$

$\mathcal{FC}$ is a *fictitious price player set*; each player $i \in \mathcal{FC}$ chooses a price $\pi_i$ from the strategy set $\mathcal{P}_i^{\mathcal{FC}} = [0, \bar{\pi}_i]$ and receives payoff

$$s_i^{\mathcal{FC}}(\pi_i; \boldsymbol{p}) = -(\pi_i - \mathcal{C}_i(\boldsymbol{p}))^2. \tag{7.9}$$

Here $\bar{\pi}_i = \sup_{\boldsymbol{p}} \mathcal{C}_i(\boldsymbol{p})$, which could be infinite for some utility functions.

In $G_{FPP}$, each user in the ad hoc network is split into two fictitious players, one in $\mathcal{FW}$ who controls power $p_i$ and the other one in $\mathcal{FC}$ who controls price $\pi_i$. Although users in the real network cooperate with each other by exchanging interference information (instead of choosing prices to maximize their surplus), each fictitious player in $G_{FPP}$ is selfish and maximizes its own payoff function. In the rest of this section, a "user" refers to one of the $M$ transmitter-receiver pairs in set $\mathcal{M}$, and a "player" refers to one of the $2M$ fictitious players in the set $\mathcal{FW} \cup \mathcal{FC}$.

In $G_{FPP}$ the players' best responses are given by

$$\mathcal{B}_i^{\mathcal{FW}}(p_{-i}, \pi_{-i}) = \mathcal{W}_i(p_{-i}, \pi_{-i}), \forall i \in \mathcal{FW}$$

and

$$\mathcal{B}_i^{\mathcal{FC}}(\boldsymbol{p}) = \mathcal{C}_i(\boldsymbol{p}), \forall i \in \mathcal{FC},$$

where $\mathcal{W}_i$ and $\mathcal{C}_i$ are the update rules for the ADP algorithm. In other words, the ADP algorithm can be interpreted as if the players in $G_{FPP}$ employ asynchronous *myopic best response (MBS)* updates, i.e., the players update their strategies according their best responses assuming the other player's strategies are fixed. It is known that the set of fixed points of MBS updates are the same as the set of NEs of a game [59, Lemma 4.2.1]. Therefore, we have:

**Lemma 7.5**  $(\boldsymbol{p}^*, \boldsymbol{\pi}^*) \in \mathcal{F}^{ADP}$ *if and only if* $(\boldsymbol{p}^*, \boldsymbol{\pi}^*)$ *is a NE of* $G_{FPP}$.

Together with Lemma 7.4, it follows that proving the convergence of asynchronous MBS updates of $G_{FPP}$ is sufficient to prove the convergence of the ADP algorithm to a solution of KKT conditions. We next analyze this convergence using supermodular game theory [59].

We first introduce some definitions.[6] A real $m$-dimensional set $\mathcal{V}$ is a *sublattice* of $\mathbb{R}^m$ if for any two elements $a, b \in \mathcal{V}$, the component-wise minimum, $a \wedge b$, and the component-wise maximum, $a \vee b$, are also in $\mathcal{V}$. In particular, a compact sublattice has a (component-wise) smallest and largest element. A twice differentiable function $f$ has *increasing differences* in variables $(x, t)$ if $\partial^2 f / \partial x \partial t \geq 0$ for any feasible $x$ and $t$.[7] A function $f$ is *supermodular* in $\boldsymbol{x} = (x_1, .., x_m)$ if it has increasing differences in $(x_i, x_j)$ for all $i \neq j$.[8] Finally, a game $G = [\mathcal{M}, \{\mathcal{P}_i\}, \{s_i\}]$ is *supermodular* if for each player $i \in \mathcal{M}$, (*a*) the strategy space $\mathcal{P}_i$ is a nonempty and compact sublattice, and (*b*) the payoff function $s_i$ is continuous in all players' strategies, is supermodular in player $i$'s own strategy, and has increasing differences between any component of player $i$'s strategy and any component of any other player's strategy. The following theorem summarizes several important properties of these games.

**Theorem 7.6**  *In a supermodular game $G = [\mathcal{M}, \{\mathcal{P}_i\}, \{s_i\}]$,*

(*a*) *The set of NEs is a nonempty and compact sublattice and so there is a component-wise smallest and largest NE.*

(*b*) *If the users' best responses are single-valued, and each user uses MBS updates starting from the smallest (largest) element of its strategy space, then the strategies monotonically converge to the smallest (largest) NE.*

(*c*) *If each user starts from any feasible strategy and uses MBS updates, the strategies will eventually lie in the set bounded component-wise by the smallest and largest NE. If the NE is unique, the MBS updates globally converge to that NE from any initial strategies.*

Properties (*a*) follows from Lemma 4.2.1 and 4.2.2 in [59]; (*b*) follows from Theorem 1 of [60] and (*c*) can be shown by Theorem 8 in [61].

Next we show that by an appropriate strategy space transformation certain instances of $G_{FPP}$ are equivalent to supermodular games, and so Theorem 7.6 applies. We first study a simple two-user network, then extend the results to an $M$-user network.

**Two-user Networks**

Let $G^2_{FPP}$ be the FPP game corresponding to a two-user network; this will be a game with four players, two in $\mathcal{FW}$ and two in $\mathcal{FC}$. First, we check whether $G^2_{FPP}$ is supermodular. Each user $i \in \mathcal{FW}$ clearly has a nonempty and compact sublattice (interval) strategy set, and so does each

---

[6]More general definitions related to supermodular games are given in [59].

[7]If we choose $x$ to maximize a twice differentiable function $f(x, t)$, then the first order condition gives $\partial f(x, t) / \partial x|_{x=x^*} = 0$, and the optimal value $x^*$ increases with $t$ if $\partial^2 f / \partial x \partial t > 0$.

[8]A function $f$ is always supermodular in a single variable $x$.

user $i \in \mathcal{FC}$ if $\bar{\pi}_i < \infty$.[9] Each player's payoff function is (trivially) supermodular in its own one-dimensional strategy space. The remaining increasing difference condition for the payoff functions does *not* hold with the original definition of strategies $(\boldsymbol{p}, \boldsymbol{\pi})$ in $G^2_{FPP}$. For example, from (7.8),

$$\frac{\partial s_i^{FW}}{\partial p_i \partial \pi_j} = -h_{ij} < 0, \forall j \neq i,$$

e.g., a higher price leads the other users to decrease their powers. However, if we define $\pi'_j = -\pi_j$ and consider an equivalent game where each user $j \in \mathcal{FC}$ chooses $\pi'_j$ from the strategy set $\left[-\bar{\pi}_j, 0\right]$, then

$$\frac{\partial s_i^{FW}}{\partial p_i \partial \pi'_j} = h_{ij} > 0, \forall j \neq i,$$

i.e., $s_i^{FW}$ has increasing differences in the strategy pair $\left(p_i, \pi'_j\right)$ (or equivalently $\left(p_j, -\pi_j\right)$). If all the users' strategies can be redefined so that each player's payoff satisfies the increasing differences property in the transformed strategies, then the *transformed FPP game* is supermodular.

Denote

$$CR_i(\gamma_i) = -\frac{\gamma_i u_i''(\gamma_i)}{u_i'(\gamma_i)},$$

and let $\gamma_i^{\min} = \min\{\gamma_i(\boldsymbol{p}) : p_i \in \mathcal{P}_i \forall i\}$ and $\gamma_i^{\max} = \max\{\gamma_i(\boldsymbol{p}) : p_i \in \mathcal{P}_i \forall i\}$. An increasing, twice continuously differentiable, and strictly concave utility function $u_i(\gamma_i)$ is defined to be

- Type I if $CR_i(\gamma_i) \in [1, 2]$ for all $\gamma_i \in \left[\gamma_i^{\min}, \gamma_i^{\max}\right]$;

- Type II if $CR_i(\gamma_i) \in (0, 1]$ for all $\gamma_i \in \left(\gamma_i^{\min}, \gamma_i^{\max}\right]$.

The term $CR_i(\gamma_i)$ is called the *coefficient of relative risk aversion* in economics and measures the relative concaveness of $u_i(\gamma_i)$. Many common utility functions are either Type I or Type II, as shown in Table 7.1.

The logarithmic utility function is both Type I and II. A Type I utility function is "more concave" than a Type II one. Namely, an increase in one user's transmission power would induce the other users to increase their powers, i.e.,

$$\frac{\partial^2 u_i(\gamma_i(\boldsymbol{p}))}{\partial p_i \partial p_j} \geq 0, \forall j \neq i.$$

A Type II utility would have the opposite effect, i.e.,

$$\frac{\partial^2 u_i(\gamma_i(\boldsymbol{p}))}{\partial p_i \partial p_j} \leq 0, \forall j \neq i.$$

---

[9]When $P_i^{\min} = 0$, this bounded price restriction is not satisfied for utilities such as $u_i(\gamma_i) = \theta_i \gamma_i^\alpha / \alpha$ with $\alpha \in [-1, 0)$, since $\pi_i = \theta_i \gamma_i^{\alpha+1} / (p_i h_{ii} B)$ is not bounded as $p_i \to 0$. However, as noted above, we can set $P_i^{\min}$ to some arbitrarily small value without affecting the performance.

**Table 7.1:** Examples of Type I and II utility functions

| Type I | Type II |
|---|---|
| $\theta_i \log(\gamma_i)$ | $\theta_i \log(\gamma_i)$ |
| $\theta_i \gamma_i^\alpha / \alpha$  (with  $\alpha \in [-1, 0)$) | $\theta_i \gamma_i^\alpha / \alpha$  (with  $\alpha \in (0, 1)$) |
| $1 - e^{-\theta_i \gamma_i}$ | $1 - e^{-\theta_i \gamma_i}$ |
| (with  $\frac{1}{\gamma_i^{\min}} \leq \theta_i \leq \frac{2}{\gamma_i^{\max}}$) | (with  $\theta_i \leq \frac{1}{\gamma_i^{\max}}$) |
| $a(\gamma_i)^2 + b\gamma_i + c$ | $a(\gamma_i)^2 + b\gamma_i + c$ |
| (with  $0 \leq -3a\gamma_i^{\max} \leq b \leq -4a\gamma_i^{\min}$) | (with  $b \geq -4a\gamma_i^{\max} > 0$) |
| | $\theta_i \log(1 + \gamma_i)$ |

The strategy spaces must be redefined in different ways for these two types of utility functions to satisfy the requirements of a supermodular game.

**Proposition 7.7**    $G_{FPP}^2$ *is supermodular in the transformed strategies* $(p_1, p_2, -\pi_1, -\pi_2)$ *if both users have Type I utility functions.*

**Proposition 7.8**    $G_{FPP}^2$ *is supermodular in the transformed strategies* $(p_1, -p_2, \pi_1, -\pi_2)$ *if both users have Type II utility functions.*

The proofs of both propositions consist of checking the increasing differences conditions for each player's payoff function. These results along with Theorem 7.6 enable us to characterize the convergence of the ADP algorithm. For example, if the two users have Type I utility functions (and $\bar{\pi}_1, \bar{\pi}_2 < \infty$), then $\mathcal{F}^{ADP}$ is nonempty. In case of multiple fixed points, there exist two extreme ones $(p^L, \pi^L)$ and $(p^R, \pi^R)$, which are the smallest and largest fixed points in terms of strategies $(p_1, p_2, -\pi_1, -\pi_2)$. If users initialize with $(p(0), \pi(0)) = (P_1^{\min}, P_2^{\min}, \bar{\pi}_1, \bar{\pi}_2)$ or $(P_1^{\max}, P_2^{\max}, 0, 0)$, the power and prices converge monotonically to $(p^L, \pi^L)$ or $(p^R, \pi^R)$, respectively. If users start from arbitrary initial power and prices, then the strategies will eventually lie in the space bounded by $(p^L, \pi^L)$ and $(p^R, \pi^R)$. Similar arguments can be made with Type II utility functions with a different strategy transformation. Convergence of the powers for both types of utilities is illustrated in Fig. 7.6.

**$M$-user Networks**

Proposition 7.7 can be easily generalized to a network with $M > 2$:

**Corollary 7.9**    *For an $M$-user network if all users have Type I utilities, $G_{FPP}$ is a supermodular in the transformed strategies* $(p, -\pi)$.

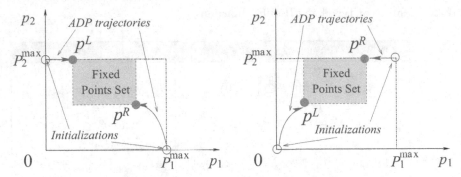

**Figure 7.6:** Examples of the trajectories of the power profiles under the ADP algorithm for a two-user network with Type I (left) or Type II (right) utility functions. In both cases, from the indicated initializations the power profiles will monotonically converge to the indicated "corner" fixed points.

In this case, Theorem 7.6 can again be used to characterize the structure of $\mathcal{F}^{ADP}$ as well as the convergence of the ADP algorithm. On the other hand, it can be seen that the strategy redefinition used in Proposition 7.8, cannot be applied with $M > 2$ users so that the increasing differences property holds for every pair of users.

With logarithmic utility functions, it is shown in [62] that Problem P1 is a strictly concave maximization problem over the transformed variables $y_i = \log p_i$. In this case Problem P1 has a unique optimal solution, which is the only point satisfying the KKT conditions. It follows from Lemma 7.4 and Lemma 7.5 that $G_{FPP}$ will have a unique NE corresponding to this optimal solution and the ADP algorithm will converge to this point from any initial choice of powers and prices.[10] With some minor additional conditions, the next proposition states that these properties generalize to other Type I utility functions.

**Proposition 7.10**    *In an $M$-user network, if for all $i \in \mathcal{M}$:*

a) $P_i^{\min} > 0$, *and*

b) $CR_i(\gamma_i) \in [a, b]$ *for all* $\gamma_i \in [\gamma_i^{\min}, \gamma_i^{\max}]$, *where* $[a, b]$ *is a strict subset of* $[1, 2]$,

*then Problem P1 has a unique optimal solution, to which the ADP algorithm globally converges.*

## 7.2.4    NUMERICAL RESULTS

We simulate a network contained in a 10m×10m square area. Transmitters are randomly placed in this area according to a uniform distribution, and the corresponding receiver is randomly placed

---

[10]Moreover, if each user $i \in \mathcal{M}$ starts from profile $(p_i(0), \pi_i(0)) = \left(P_i^{\min}, \theta_i/(n_0 B)\right)$ or $(P_i^{\max}, 0)$, then their strategies will monotonically converge to this fixed point.

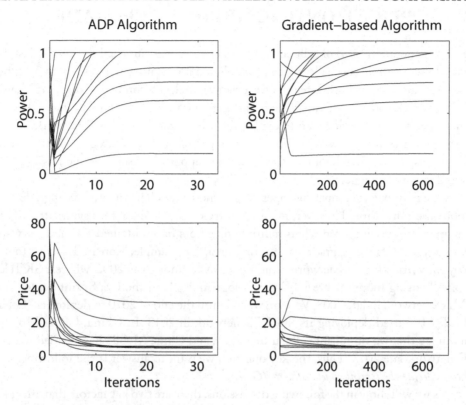

**Figure 7.7:** Convergence of the prices and powers for the ADP algorithm (left) and a gradient algorithm (right) in a network with 10 users and logarithmic utility functions. Each curve corresponds to the power or price for one user with a random initialization.

within 6m×6m square centered around the transmitter. There are $M = 10$ users, each with utility $u_i = \log(\gamma_i)$. The channel gains $h_{ij} = d_{ij}^{-4}$, $P_i^{\max}/n_0{=}40dB$, and $B{=}128$MHz. Figure 7.7 shows the convergence of the powers and prices for each user under the ADP algorithm for a typical realization, starting from random initializations. Also, for comparison we show the convergence of these quantities using a gradient-based algorithm as in [62] with a step-size of 0.01.[11] Both algorithms converge to the optimal power allocation, but the ADP algorithm converges much faster; in all the cases we have simulated, the ADP algorithm converges about 10 times faster than the gradient-based algorithm (if the latter converges). The ADP algorithm, by adapting power according to the best response updates, is essentially using an "adaptive step-size" algorithm: users adapt the power in "larger" step-sizes when they are far away from the optimal solution, and use finer steps when close to the optimal.

---

[11]In our experiments, a larger step-size than 0.01 would often not converge.

## 7.3    APPLICATION II: 4G NETWORK UPGRADE

The third generation (3G) of cellular wireless networks was launched during the last decade. It has provided users with high-quality voice channels and moderate data rates (up to 2 Mbps). However, 3G service cannot seamlessly integrate the existing wireless technologies (e.g., GSM, wireless LAN, and Bluetooth), and cannot satisfy users' fast growing needs for high data rates. Thus, most major cellular operators worldwide plan to deploy the fourth-generation (4G) networks to provide much higher data rates (up to hundreds of megabits per second) and integrate heterogeneous wireless technologies. The 4G technology is expected to support new services such as high-quality video chat and video conferencing.

One may expect competitive operators in the same cellular market to upgrade to a 4G service at about the same time. However, many industry examples show that symmetric 4G upgrades do not happen in practice, even when multiple operators have obtained the necessary spectrum and technology patents for upgrade. In South Korea, for example, Korean Telecom took the lead to deploy the world's first 4G network using WiMAX technology in 2006, whereas SK Telecom started to upgrade using more mature LTE technology in 2011. In the US, Sprint deployed the first 4G WiMAX network in late 2008, Verizon waited until the end of 2010 to deploy its 4G LTE network, and AT&T started deploying its 4G LTE network in 2013. In China, China Mobile and China Unicom are the two dominant cellular operators, and China Mobile has decided to first deploy 4G LTE network late 2013. Thus, the key question we want to answer is the following: *how do cellular operators decide the timing to upgrade to 4G networks?*

As we will show in the following discussions, there are two key factors that affect the operators' upgrade decisions: namely, *4G upgrade cost* and *user switching cost*. An existing 3G user can switch to the 4G service of the same operator or of a different operator, depending on how large the switching cost is. In a monopoly market where only a dominant operator can choose to upgrade to 4G, this operator can use the 4G service to capture a larger market share from small operators. The discussions about the more interesting competition market can be found in [7].

### 7.3.1    SYSTEM MODEL

**Value of Cellular Networks**

Here we adopt the $N \log(N)$ Law of the network effect [57], where the network value with $N$ users is proportional to $N \log(N)$. The operator of a cellular network prefers a large network value; this is because the revenue he obtains by charging users can be proportional to the network value. Notice that the value of a 4G network is larger than a 3G network even when two networks have the same number of users. This is because the communication between two 4G users is more efficient and more frequent than between two 3G users. Because the average data rate in the 4G service is five to ten times faster than the 3G, a 4G network can support many new applications. We denote the efficiency ratio between 3G and 4G services as $\gamma \in (0, 1)$. That is, by serving all his users via QoS-guaranteed 4G rather than 3G services, an operator obtains a larger (normalized) revenue

$N \log(N)$ instead of $\gamma N \log(N)$.[12] Note that this result holds for a single operator's network that is not connected to other networks.

Next we discuss the revenues of *multiple* operators whose networks (e.g., two 3G networks) are interconnected. For the purpose of illustration, we consider two networks that contain $N_1$ and $N_2$ users, respectively. The whole market covers $N = N_1 + N_2$ users. We assume that two operators' 3G (and later 4G) services are equally good to users, and the efficiency ratio $\gamma$ is the same for both operators. The traffic between two users can be intra-network (when both users belong to the same operator) or inter-network (when two users belong to different operators), and the revenue calculations in the two cases are different. We assume that the user who originates the communication session (irrespective of whether the same network or to the other network) pays for the communication. This is motivated by the industry observations in the EU and many Asian countries. Before analyzing each operator's revenue, we first introduce two practical concepts in cellular market: "termination rate" and "user ignorance."

When two users of the same operator 1 communicate with each other, the calling user only pays operator 1. But when an operator 1's user calls an operator 2's user, operator 2 charges a *termination rate* for the incoming call.[13] We denote the two operators' revenue-sharing portion per inter-network call as $\eta$, where the value of $\eta \in (0, 1)$ depends on the agreement between the two operators or on governments' regulation on termination rate.

User ignorance is a unique problem in the wireless cellular network, where users are often not able to identify which specific network they are calling. Mobile number portability further exacerbates this problem. Thus, a typical user's evaluation of two interconnected 3G networks does not depend on which network he belongs to, and equals $\gamma \log(N)$ where $N = N_1 + N_2$. We assume a call from any user terminates at a user in network $i \in \{1, 2\}$ with a probability of $N_i/N$. The operators' revenues when they are both providing 3G services are given in Lemma 7.11.

**Lemma 7.11** *When operators 1 and 2 provide 3G services, their revenues are $\gamma N_1 \log(N)$ and $\gamma N_2 \log(N)$, respectively.*

Both operators' revenues are linear in their numbers of users (or market share), and are independent of the sharing portion $\eta$ of the inter-network revenue. Intuitively, the inter-network traffic between two networks is bidirectional: when a user originates a call from network 1 to another user in network 2, his inter-network traffic generates a fraction $\eta$ of corresponding revenue to operator 1; when the other user calls back from network 2 to network 1, he generates a fraction $1 - \eta$ of the same amount of revenue to operator 1. Thus, an operator's total revenue is independent of $\eta$. Later, in Section 7.3.2, we show that such independence on $\eta$ also applies when the two operators both provide 4G services or provide mixed 3G and 4G services.

---

[12]We assume that an operator's operational cost (proportional to network value) has been deducted already, and thus the revenue here is the normalized one.

[13]In the US, termination rate follows "Bill and Keep" and is low. Then operator 1 can keep most of the calling user's payment. In EU, however, termination rate follows "Calling Party Pays" and is much higher. Then most of the calling user's payment to operator 1 is used to compensate for the termination rate charged by operator 2.

### User Churn during Upgrade from 3G to 4G Services

When 4G service becomes available in the market (offered by one or both networks), the existing 3G users have an incentive to switch to the new service to experience a better QoS. Such user churn does not happen simultaneously for all users, because different users have different sensitivities to quality improvements and switching costs. We use two parameters $\lambda$ and $\alpha$ to model the user churn within and between operators:

- *Intra-network user churn:* If an operator provides 4G in addition to his existing 3G service, his 3G users need to buy new mobile phones to use the 4G service. The users also spend time to learn how to use the 4G service on their new phones. We use $\lambda$ to denote the users' switching rate to the 4G service within the same network.

- *Inter-network user churn:* If a 3G user wants to switch to another network's 4G service, he either waits till his current 3G contract expires, or pays for the penalty of immediate contract termination. This means that inter-network user churn incurs an additional cost on top of the mobile device update, and thus the switching rate will be smaller than the intra-network user churn. We use $\alpha\lambda$ to denote the users' inter-network switching rate to 4G service, where $\alpha \in (0, 1)$ reflects the transaction cost of switching operators.

We illustrate the process of user churn through a continuous time model. The starting time $t = 0$ denotes the time when the spectrum resource and the 4G technology are available for at least one operator. We also assume that the portion of users switching to the 4G service follows the exponential distribution (at rate $\lambda$ for intra-network churn and $\alpha\lambda$ for inter-network churn).

As an example, assume that operator 1 introduces a 4G service at time $t = T_1$ while operator 2 decides not to upgrade. The numbers of operator 1's 4G users and 3G users at any time $t \geq 0$ are $N_1^{4G}(t)$ and $N_1^{3G}(t)$, respectively. The number of operator 2's 3G users at time $t \geq 0$ is $N_2^{3G}(t)$. As time $t$ increases (from $T_1$), 3G users in both networks start to churn to 4G service, and $\forall t \geq 0$

$$N_1^{3G}(t) = N_1 e^{-\lambda \cdot \max(t-T_1,0)}, \ N_2^{3G}(t) = N_2 e^{-\alpha\lambda \cdot \max(t-T_1,0)}, \tag{7.10}$$

and operator 1's 4G service gains an increasing market share,

$$N_1^{4G}(t) = N - N_1 e^{-\lambda \cdot \max(t-T_1,0)} - N_2 e^{-\alpha\lambda \cdot \max(t-T_1,0)}. \tag{7.11}$$

We illustrate (7.10) and (7.11) in Fig. 7.8. We can see that operator 1's early upgrade attracts users from his competitor and increases his market share. Notice that (7.11) increases with $\alpha$, thus operator 1 captures a large market share when $\alpha$ is large (i.e., the switching cost is low).

### Operators' Revenues and Upgrade Costs

Because of the time discount, an operator values the current revenue more than the same amount of revenue in the future. We denote the discount rate over time as $S$, and the discount factor is thus $e^{-St}$ at time $t$.

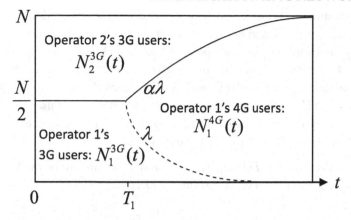

**Figure 7.8:** The numbers of users in the operators' different services as functions of time $t$. Here, operator 1 upgrades at $T_1$ and operator 2 does not upgrade.

We approximate one operator's 4G upgrade cost as a one-time investment. This is a practical approximation, as an operator's initial investment of wireless spectrum and infrastructure can be much higher than the maintenance costs in the future. For example, spectrum is a very scarce resource that is allocated (auctioned) infrequently by government agencies. Thus, an operator cannot obtain additional spectrum frequently after his 4G upgrade. To ensure a good initial 4G coverage, an operator also needs to update many base stations to cover at least a whole city all at once. Otherwise, 4G users would be unhappy with the service, and this would damage the operator's reputation. That is why Sprint and Verizon covered many markets in their initial launch of their 4G services.

More specifically, we denote the 4G upgrade cost at $t = 0$ as $K$, which discounts over time at a rate $U$. Thus, if an operator upgrades at time $t$, he needs to pay an upgrade cost $Ke^{-Ut}$. We should point out that the upgrade cost decreases faster than the normal discount rate (i.e., $U > S$). This happens because the upgrade cost decreases due to both technology improvement and time discount.

Based on these discussions on revenue and upgrade cost, we define an operator's profit as the difference between his revenue in the long run and the one-time upgrade cost. Without loss of generality, we will normalize an operator's revenue rate (at any time $t$), total revenue, and upgrade cost by $N \log(N)$, where $N$ is the total number of users in the market.

## 7.3.2   4G MONOPOLY MARKET

As an illustration, we will look at the case where only operator 1 can choose to upgrade from 3G to 4G, while the other operators (one or more) always offer the 3G service because of the lack of financial resources or the necessary technology. This can be a reasonable model, for example, for countries such as Mexico and some Latin American ones, where America Movil is the dominant cellular operator in the 3G market. As the world's fourth-largest cellular network operator, America

Movil has the advantage over other small local operators in winning additional spectrum via auctions and obtaining LTE patents, and he is expected to be the 4G monopolist in that area.

The key question here is how operator 1 should choose his upgrade time $T_1$ from the 3G service to the 4G service. $T_1 = 0$ means that operator 1 upgrades at the earliest time that the spectrum and technology are available, and $T_1 > 0$ means that operator 1 chooses to upgrade later to take advantage of the reduction in the upgrade cost. Because of user churns from the 3G to the 4G service, the operators' market shares and revenue rates change after time $T_1$. For that reason, we analyze time periods $t \leq T_1$ and $t > T_1$ separately.

- *Before 4G upgrade ($t \leq T_1$):* Operator 1's and other operators' market shares do not change over time. Operator 1's revenue rate at time $t$ is

$$\pi_1^{3G-3G}(t) = \gamma \frac{N_1}{N},$$

which is independent of time $t$. His revenue during this time period is

$$\pi_{1,t \leq T_1}^{3G-3G} = \int_0^{T_1} \pi_1^{3G-3G}(t)e^{-St}dt = \frac{\gamma N_1}{SN}(1 - e^{-ST_1}). \tag{7.12}$$

- *After 4G upgrade ($t > T_1$):* Operator 1's market share increases over time, and the other operators' total market share (denoted by $N_2^{3G}(t)/N$) decreases over time. We denote operator 1's numbers of 3G users and 4G users as $N_1^{3G}(t)$ and $N_1^{4G}(t)$, respectively, and we have $N_1^{3G}(t) + N_1^{4G}(t) + N_2^{3G}(t) = N$. This implies that

$$N_2^{3G}(t) = (N - N_1)e^{-\alpha\lambda(t-T_1)}, N_1^{3G}(t) = N_1 e^{-\lambda(t-T_1)},$$

and

$$N_1^{4G}(t) = N - N_1 e^{-\lambda(t-T_1)} - (N - N_1)e^{-\alpha\lambda(t-T_1)}.$$

Note that a 3G user's communication with a 3G or a 4G user is still based on the 3G standard, and only the communication between two 4G users can achieve a high 4G standard QoS. Operator 1's revenue rate is

$$\pi_1^{4G-3G}(t) = \frac{\gamma N_1^{3G}(t)}{N} + \frac{N_1^{4G}(t)}{N}\left(\frac{N_1^{4G}(t) + \gamma N_1^{3G}(t)}{N} + \frac{\gamma N_2^{3G}(t)}{N}\right),$$

which is independent of the revenue sharing ratio $\eta$ between the calling party and receiving party. Operator 1's revenue during this time period is then

$$\pi_{1,t>T_1}^{4G-3G} = \int_{T_1}^{\infty} \pi_1^{4G-3G}(t)e^{-St}dt, = e^{-ST_1}\left(\frac{1}{S} + (1 - \gamma)\left(\frac{N_1}{N}\right)^2 \frac{1}{2\lambda + S}\right), \tag{7.13}$$

where $t \to \infty$ is an approximation of the long-term 4G service provision (e.g., one decade) before the emergence of the next generation standard. This approximation is reasonable since the revenue in the distant future becomes less important because of the discount.

Figure 7.8 illustrates how the numbers of users of operators' different services change over time. Before operator 1's upgrade (e.g., $t \leq T_1$ in Fig. 7.8), the number of total users in each network does not change; after operator 1's upgrade, operator 1's and the other operators' 3G users switch to the new 4G service at rates $\lambda$ and $\alpha\lambda$, respectively.

By considering (7.12), (7.13), and the decreasing cost $Ke^{-UT_1}$, operator 1's long-term profit when choosing an upgrade time $T_1$ is

$$
\begin{aligned}
\pi_1(T_1) =& \pi_{1,t \leq T_1}^{3G-3G} + \pi_{1,t>T_1}^{4G-3G} - Ke^{-UT_1} \\
=& e^{-ST_1} \left( \frac{1}{S} + (1-\gamma)\frac{\left(\frac{N_1}{N}\right)^2}{2\lambda+S} + (1-\gamma)\frac{\left(\frac{N-N_1}{N}\right)^2}{2\alpha\lambda+S} \right) \\
& - e^{-ST_1} \left( 2(1-\gamma)\frac{\frac{N_1}{N}}{\lambda+S} + (2-\gamma)\frac{\frac{N-N_1}{N}}{\alpha\lambda+S} \right) - Ke^{-UT_1} \\
& + 2e^{-ST_1}(1-\gamma)\frac{\frac{N_1(N-N_1)}{N^2}}{(1+\alpha)\lambda+S} + \frac{N_1\gamma}{NS}(1-e^{-ST_1}).
\end{aligned}
\tag{7.14}
$$

We can show that $\pi_1(T_1)$ in (7.14) is strictly concave in $T_1$, thus we can compute the optimal upgrade time $T_1^*$ by solving the first-order condition. The optimal upgrade time depends on the following upgrade cost threshold in the monopoly 4G market,

$$
\begin{aligned}
K_{th}^{mono} =& (1-\gamma)\frac{\frac{\left(\frac{N_1}{N}\right)^2}{2\lambda+S} + \frac{\left(\frac{N-N_1}{N}\right)^2}{2\alpha\lambda+S} - \frac{2\frac{N_1}{N}}{\lambda+S} + \frac{2\frac{N_1(N-N_1)}{N^2}}{(1+\alpha)\lambda+S}}{U/S} \\
& + \frac{1 - \gamma\frac{N_1}{N} - (2-\gamma)\frac{N-N_1}{N}\frac{S}{\alpha\lambda+S}}{U}.
\end{aligned}
\tag{7.15}
$$

**Theorem 7.12**  *Operator 1's optimal upgrade time in a 4G monopoly market is:*

- *Low cost regime (upgrade cost $K \leq K_{th}^{mono}$): operator 1 upgrades at $T_1^* = 0$.*

- *High cost regime ($K > K_{th}^{mono}$): operator 1 upgrades at*

$$
T_1^* = \frac{1}{U-S} \log\left(\frac{K}{K_{th}^{mono}}\right) > 0.
\tag{7.16}
$$

Intuitively, an early upgrade gives operator 1 a larger market share and enables him to get a higher revenue via the more efficient 4G service. Such advantage is especially obvious in the low cost regime where the upgrade cost $K$ is small.

## 7.4    CHAPTER SUMMARY

This chapter discussed how we should deal with negative and positive externalities in networks. We first explained the concept of externality, which reflects the side effect that is imposed by the actions of a player on a third party not directly involved. We then illustrated that both positive and negative externalities can lead to a deviation from the social optimal solution. As an example of the negative externality, we illustrated how pollution from a chemical plan will affect the business of a water company, and how such negative externality can be internalized by the method of Pigovian tax. An example of the positive externality, we explained the different kinds of network effects.

We then illustrated the application of theory using two applications. In the first application, we considered the problem of optimal distributed power control in wireless ad hoc networks. The mutual interferences pose negative externality among wireless users. To mitigate such negative externality, the wireless users will charge each other the interference prices, which are basically distributively computed Pigovian tax. Under proper technical conditions, the ADP algorithm designed based on the asynchronous power and price updates will converge to a global or local optimal solution, much faster than the usual gradient based method with small step sizes. In the second application, we considered the economics of 4G upgrade, where the cellular operators need to consider the impact of network grade in terms of updating cost and benefits of user switching. The positive network effect will induce a monopoly operator to upgrade as early as possible when the upgrading cost is small. For more details especially mathematical proofs related to the two applications, please see [7, 63].

## 7.5    EXERCISES

1. Suppose each pack of cigarette smoked creates $X$ units of externality experienced by other members of society. Such an externality may include cigarette smokers' excess use of public health services, the medical costs of the second-hand smokers, etc. Explain how a Pigovian tax can be used to correct the externality (such that the smokers consume fewer cigarettes than socially desired). Also find another way that the government can use to correct the over-consumption of cigarettes due to neglect of the negative externality.

2. Consider $N$ ordered potential customers for a network product, where the $i$th customer is willing to pay $i \log(n)$ dollars for the product if there are a total of $n$ customers using the product. Assume that the product is charged at a price $p$. Compute the number of customers who are willing to purchase the product at the "equilibrium," where there is no customer willing to change his decision.

# CHAPTER 8

# Outlook

In this book, we have assumed that all market players have access to *complete market information*. We have also assumed that *one player (or one type of player) has the market power* to determine the key market parameters (e.g., production quality or price), and the other market players can only choose to accept the arrangements or reject to participate in the market. Unfortunately, these two assumptions are often violated in practice. Usually market information is incomplete to most market players, and the market power is distributed among various market players.

A number of new issues arise under information asymmetry, among which the most important one is the *truthfulness* (also called *incentive compatibility*). That is, how to design a truthful (or incentive compatible) mechanism that credibly elicits the private information held by some market players. The significance of the truthful mechanism is suggested by the revelation principle [64], which states that "for any outcome resulting from any mechanism, there always exists a *payoff-equivalent* revelation mechanism where the players truthfully report their private information." The principle is extremely powerful. It allows a market designer to solve for an outcome or equilibrium by assuming all players truthfully report their private information (subject to the incentive compatibility constraint). This means that to find an optimal mechanism to achieve a certain objective (e.g., profit maximization or social welfare maximization), we do not need to search from the infinitely large set of mechanisms where players can act arbitrarily, but only need to consider those truthful mechanisms where players act truthfully. Typical examples of truthful mechanisms include auction and contract.[1]

Problems become significantly different when the market power is distributed among multiple market players. Specifically, when one player or one type of player (e.g., firms or consumers) has the total market power, the player(s) will try to extract the social surplus as much as possible. For example, in a monopoly market where the monopolist has the total market power, the monopolist can set a monopoly price or perform price discrimination to maximize his profit. In an oligopoly market where the firms have the total market power, they can set strategic prices or quantities to maximize their own profits against others' strategies. This type of self-interested interaction among players is essentially referred to as the non-cooperative game theory. When the market power is distributed among different types of market player (e.g., firms and consumers), such self-interested interactions may no longer work due to the conflicting interests among the players. Consider a simple example where a firm sells a single product to a consumer. How to determine the price of the product if both the firm and the consumer have certain market power? This problem cannot be solved by a self-interested interaction, since the increase of one player's surplus must lead to the

---

[1] Note this is not to say that all auction designs or contract designs are truthful. But rather, a considerable part (and possibly, the most important part) of auctions and contracts are truthful mechanisms.

decrease of the other's surplus. A well-studied discipline to this type of problem is *bargaining*. Simply speaking, bargaining solution is such an outcome that both players feel *acceptable*, rather than strictly prefer in terms of certain criterion. This type of bargaining interaction is essentially referred to as the cooperative game theory.

In the rest of the chapter, we will briefly discuss the connections and differences between pricing, auction, contract, and bargaining models, and provide pointers to some key related papers in the wireless literature (without an intention of being exhaustive). We hope to provide more detailed discussions regarding the theories and applications of these different economic mechanisms in a future book.

## 8.1  AUCTION

An auction is a process of buying and selling goods or services by offering them up for bid, taking bids, and then selling the item(s) to the highest bidder(s). Typical issues studied by auction theorists include the efficiency of a given auction design, optimal and equilibrium bidding strategies, and revenue comparison. There are many possible designs (or sets of rules) for an auction, among which the most important two are the *allocation rule* (who is/are the winner(s) of an auction) and the *payment rule* (what will be the payment(s) of the winner(s)). In this sense, an auction can be viewed as a special kind of the pricing model.

One key difference between the pricing model and the auction model is their application scenarios. Specifically, the network pricing introduced in this book is often used in the symmetric and complete information scenario, where the decision-makers know complete information about the commodities or the market; whereas the auction model is often used in the asymmetric information scenario where the market players (called bidders) hold certain private information that the decision-makers (called auctioneers) do not know. In a pricing model, the decision-makers determine the market price based on their known information; while in an auction model, the auctioneers let the market (i.e., the community of bidders) set the price, and account for the uncertainty about the bidders' valuations. With a careful design, the bidders have the incentive to bid for the commodity in a truthful manner, and the auctioneers can efficiently allocate the commodity without knowing the bidders' private valuations in advance.

Auction has been widely used for wireless network resource allocation and performance optimization. In [65], Huang et al. proposed two divisible auction mechanisms for power allocation to achieve efficiency and fairness, respectively. In [66], Huang et al. extended the auction mechanisms to a cooperative communication system with multiple relays. In [67], Li et al. proposed several truthful (strategy-proof) spectrum auction mechanisms to achieve the efficiency closed to social optimal. In [68], Gandhi et al. proposed a real-time spectrum auction framework under interference constraints. In [69, 70], Zheng et al. proposed truthful single side spectrum auction and double spectrum auction, respectively, both considering spectrum reuse. In [71], Wang et al. proposed a general framework for truthful double auction for spectrum sharing. In [72], Gao et al. proposed a multi-shot spectrum auction mechanism to achieve social optimal efficiency in dynamic spectrum sharing.

## 8.2    CONTRACT

A contract is an agreement entered into voluntarily by two or more parties with the intention of creating a legal obligation. In economics, contract theory studies how the economic agents construct *contractual* arrangements, generally in the presence of asymmetric information. Thus, it is closely connected to the truthful (or incentive compatible) mechanism design. Several well known contract models include *moral hazard*, *adverse selection*, *signaling*, and *screening*. The common spirit of these models is to motivate one party (the agent) to act in the best interests of another (the principal).

In moral hazard models, the information asymmetry is generated by the principal's inability to observe and/or verify the agent's action (termed as hidden action). Contracts that depend on observable and verifiable output can often be employed to create incentives for the agent to act in the principal's interest. In adverse selection models, the principal is not informed about a certain characteristic of the agent (termed as hidden information). Two commonly used methods to model adverse selection are signaling games and screening games. The idea of signaling is that one party (usually the agent) credibly conveys some information about itself to another party (usually the principal). The idea of screening is that one party (usually the principal) offers multiple contract options, which are incentive compatible for another party (usually the agent) such that every agent selects the option intended for his type. The main difference between signaling and screening is who moves first. In signaling games, the informed agent moves (signaling) first, and the process is essentially a Stackelberg game with the agent as the leader. In screening games, however, the uninformed principal moves (offering the options) first, and the process is essentially a Stackelberg game with the principal as the leader. In the context of monopoly market, the second degree price discrimination (in Section 5) is essentially a screening model.

Contract has also been widely used in wireless networks. In [73], Gao et al. proposed a quality-price screening contract for secondary spectrum trading, where the seller offers a menu of prices for different qualities to attract different types of buyers. The authors showed that with the contract, the seller can extract more surplus from the buyers with private information. In [74], Kalathil et al. proposed a contract-based spectrum-sharing mechanism to avoid possible manipulating in a spectrum auction. The authors showed that it is possible to achieve socially optimal rate allocations with contracts in licensed bands. In [75], Duan et al. proposed a time-power screening contract for cooperative spectrum sharing between primary users and secondary users, where the secondary users relay traffic for PUs in exchange for the guaranteed access time on the primary users' licensed spectrums. In [76, 77], Kasbekar and Sarkar et al. considered the secondary spectrum trading with two types of contracts: the guaranteed-bandwidth contract, which provides guaranteed access to a certain amount of bandwidth for a specified duration of time, and the opportunistic-access contract, which offers restricted (uncertain) access rights on a certain amount of bandwidth at the current time slot. In [78], Gao et al. studies the secondary spectrum trading in a hybrid market with both contract users and spot purchasing users.

## 8.3    BARGAINING

Bargaining is a type of negotiation in which the buyer and seller of a good or service discuss the price which will be paid and the exact nature of the transaction that will take place, and eventually come to an agreement. In this sense, bargaining is an alternative pricing strategy to fixed prices. Bargaining arises when the market power is distributed among different market players (and thus no participant has the total market power to determine the solution solely). Solutions to bargaining come in two flavors: an *axiomatic approach* where desired properties of a solution are satisfied, and a *strategic approach* where the bargaining procedure is modeled in detail as a sequential game. Typical solutions of bargaining include the Nash bargaining solution, Shapely value, Harsanyi value, and so on.

The study of bargaining was initiated by J. Nash in 1950 [79], who provided an axiomatic solution for the outcome of the negotiation among two players. In 1982, A. Rubinstein proposed a sequential non-cooperative game between two players [80], where the player alternate offers through an infinite time horizon. As one player offers a proposal, the other player decides to accept or reject. If a proposal is rejected (by the responder), the proposer and responder change their roles, that is, the previous responder becomes a proposer offering a new proposal, and the previous proposer becomes a responder deciding to accept or reject the new proposal. Rubinstein characterized the subgame perfect equilibrium of this game, and concluded that the subgame perfect equilibrium of this non-cooperative Rubinstein bargaining game is closely related to the Nash bargaining solution given by the axiomatic approach.

Although many studies considered that players bargain independently and in an uncoordinated fashion, a survey of recent economic journals reveals that most applied bargaining papers actually analyze group bargaining problems [81]. That is, more often than not, players form groups and bargain jointly in order to improve their anticipated payoff. Examples include labor disputes between the management which represents the stockholders of a factory, and a union which represents the workers [82]. In order to predict the bargaining result in such settings, it is necessary to analyze both the inter-group bargaining and the intra-group bargaining. Usually, bargaining first takes place among the different groups, and then members within each group bargain with each other in order to distribute the acquired welfare. In most cases, the grouping improves the payoff of the group members [83, 84], since it leverages their bargaining power. Moreover, often the bargaining outcome depends on the bargaining protocol, i.e., bargaining concurrently or sequentially (and the sequence the players bargain). This aspect was studied in [85, 86], where one dominant player optimally selects a weak player to bargain in each stage.

Bargaining has also been widely used in wireless networks. In [87], Zhang et al. proposed a cooperation bandwidth allocation strategy based on the Nash bargaining solution in a wireless cooperative relaying network. In [88], Cao et al. proposed a local bargaining approach to fair spectrum allocation in mobile ad-hoc networks. In [89], Han et al. proposed a fair scheme to allocate subcarrier, rate, and power for multiuser OFDMA systems based on Nash bargaining solutions and coalitions. The above works studied the bargaining problem using axiomatic approaches. In [90], Yan et al. studied the bargaining problem using strategic approaches. Specifically, they considered dynamic bargaining between one primary user and several secondary users in a cognitive cooperative network with incomplete network information. Moreover, in [91], Boche et al. studied the necessary requirements for the existence and uniqueness of Nash bargaining solution and proportional fairness solution.

# Bibliography

[1] G. Staple and K. Werbach, "The end of spectrum scarcity," *IEEE Spectrum*, pp. 48–52, March 2004. DOI: 10.1109/MSPEC.2004.1270548. 2

[2] M. McHenry, "NSF spectrum occupancy measurements project summary," *Shared Spectrum Company*, 2005. 2

[3] P. Bahl, R. Chandra, T. Moscibroda, R. Murty, and M. Welsh, "White space networking with Wi-Fi like connectivity," *ACM SIGCOMM Computer Communication Review*, vol. 39, no. 4, pp. 27–38, 2009. DOI: 10.1145/1594977.1592573. 2

[4] A. Technica, "Broadcasters sue FCC over white space broadband decision." [Online]. Available: http://arstechnica.com/tech-policy/news/2009/03/broadcasters-sue-fcc-over-white-space-broadband-decision.ars 2

[5] T. Nguyen, H. Zhou, R. Berry, M. Honig, and R. Vohra, "The impact of additional unlicensed spectrum on wireless services competition," in *IEEE Symposium on New Frontiers in Dynamic Spectrum Access Networks (DySPAN)*, 2011. 2

[6] Cisco, "Cisco visual networking index: Global mobile data traffic forecast update, 2011–2016," White Paper, 2012.

[7] L. Duan, J. Huang, and J. Walrand, "Economic analysis of 4G network upgrade," in *IEEE INFOCOM*, Turin, April 2013. 4, 138, 144

[8] T. S. Rappaport *et al.*, *Wireless communications: principles and practice.* Prentice Hall, 1996, vol. 2. 7, 17

[9] A. Goldsmith, *Wireless communications.* Cambridge University Press, 2005. 7, 17

[10] D. Tse and P. Viswanath, *Fundamentals of wireless communication.* Cambridge University Press, 2005. 7, 17

[11] V. Garg, *Wireless Communications & Networking.* Morgan Kaufmann, 2010. 7, 17

[12] Wikipedia, "Radio propagation." [Online]. Available: http://en.wikipedia.org/wiki/Radio_propagation 7, 17

[13] Wikipedia, "Communication channel." [Online]. Available: http://en.wikipedia.org/wiki/Communication_channel 7, 17

[14] Wikipedia, "Multiplexing." [Online]. Available: http://en.wikipedia.org/wiki/Multiplexing 7, 17

[15] Wikipedia, "Cognitive radio." [Online]. Available: http://en.wikipedia.org/wiki/Cognitive_radio 7, 17

[16] B. P. Pashigian, *Price Theory and Applications.* McGraw-Hill Inc., 1995. 19, 61

[17] S. Landsburg, *Price Theory and Applications.* Cengage Learning, 2010. 19, 61, 63, 65, 66, 67, 69, 83

[18] A. Mas-Colell, M. Whinston, and J. Green, *Microeconomic theory.* Oxford University Press, 1995. 19, 105

[19] S. Boyd and L. Vanderberghe, *Convex Optimization.* Cambridge University Press, 2004. 31, 40, 45

[20] D. Bertsekas, *Nonlinear Programming*, 2nd ed., Belmont, Massachusetts: Athena Scientific, 1999. 31, 38, 45, 54, 129

[21] K. Arrow, L. Hurwicz, and H. Uzawa, "Constraint qualifications in maximization problems," *Naval Research Logistics Quarterly*, vol. 8, no. 2, pp. 175–191, 1961. DOI: 10.1002/nav.3800080206. 42

[22] H. Kuhn and A. Tucker, "Nonlinear programming," in *Berkeley Symposium on Mathematical Statistics and Probability.* California, 1951. 43

[23] J. Huang, Z. Li, M. Chiang, and A. K. Katsaggelos, "Joint source adaptation and resource pricing for multi-user wireless video streaming," *IEEE Transactions on Circuits and Systems for Video Technology*, vol. 18, no. 5, pp. 582–595, May 2008. DOI: 10.1109/TCSVT.2008.919109. 47, 51, 59

[24] T. M. Cover and J. Thomas, *Elements of Information Theory.* Wiley-Interscience, 1991. DOI: 10.1002/0471200611. 47

[25] R. Srikant, *The Mathematics of Internet Congestion Control.* Birkhauser Boston, 2004. DOI: 10.1007/978-0-8176-8216-3. 48

[26] A. Sampath, P. S. Kumar, and J. Holtzman, "Power control and resource management for multimedia CDMA wireless system," in *IEEE PIMRC*, 1995. DOI: 10.1109/PIMRC.1995.476272. 49

[27] K. Kumaran and L. Qian, "Uplink Scheduling in CDMA Packet-Data Systems," *Wireless Networks*, vol. 12, no. 1, pp. 33–43, 2006. DOI: 10.1007/s11276-006-6148-7. 50

[28] P. Marbach, "Analysis of a static pricing scheme for priority services," *IEEE/ACM Transactions on Networking*, vol. 12, no. 2, pp. 312–325, 2004. DOI: 10.1109/TNET.2004.826275. 52

[29] X. Lin and N. B. Shroff, "Utility maximization for communication networks with multi-path routing," *IEEE Transactions on Automatic Control*, vol. 51, no. 5, pp. 766–781, 2006. DOI: 10.1109/TAC.2006.875032. 54

[30] T. Voice, "Stability of congestion control algorithms with multi-path routing and linear stochastic modelling of congestion control," Ph.D. dissertation, University of Cambridge, Cambridge, UK, 2006. 54

[31] V. Gajic, J. Huang, and B. Rimoldi, "Competition of wireless providers for atomic users," *IEEE/ACM Transactions on Networking*, forthcoming. DOI: 10.1109/TNET.2013.2255889. 54, 56, 59, 105, 116

[32] M. Chen and J. Huang, "Optimal resource allocation for OFDM uplink communication: A primal-dual approach," in *Conference on Information Sciences and Systems*, Princeton University, NJ, USA, March 2008, pp. 926–931. DOI: 10.1109/CISS.2008.4558651. 55

[33] L. Duan, J. Huang, and B. Shou, "Economics of femtocell service provision," *IEEE Transactions on Mobile Computing*, forthcoming. DOI: 10.1109/TMC.2012.193. 72, 83, 84

[34] S. Li and J. Huang, "Revenue maximization for communication networks with usage-based pricing," *IEEE/ACM Transactions on Networking*, forthcoming. 72, 76, 80, 84

[35] V. Valancius, C. Lumezanu, N. Feamster, R. Johari, and V. Vazirani, "How many tiers? pricing in the internet transit market," *SIGCOMM-Computer Communication Review*, vol. 41, no. 4, p. 194, 2011. DOI: 10.1145/2043164.2018459. 78

[36] J. van Lint and R. Wilson, *A Course in Combinatorics*.   Cambridge University Press, 2001. 80

[37] B. Pashigian, *Price theory and applications*.   McGraw-Hill, 1998. 83

[38] M. Osborne and A. Rubinstein, *A course in game theory*.   The MIT Press, 1994. 87, 92, 93, 94, 99, 101, 103

[39] R. Gibbons, *A Primer in Game Theory*.   Prentice Hall, 1992. 87, 92, 93, 94, 99, 101, 103

[40] J. Friedman, *Game theory with applications to economics*.   Oxford University Press, 1986. 87

[41] L. Petrosjan and V. Mazalov, *Game theory and applications*.   Nova Publishers, 2002, vol. 8. 87

[42] S. Kakutani, "A generalization of brouwer's fixed point theorem," *Duke Mathematical Journal*, vol. 8, no. 3, pp. 457–459, 1941. DOI: 10.1215/S0012-7094-41-00838-4. 93

[43] I. Menache and A. Ozdaglar, "Network games: Theory, models, and dynamics," *Synthesis Lectures on Communication Networks*, vol. 4, no. 1, pp. 1–159, 2011. DOI: 10.2200/S00330ED1V01Y201101CNT009. 94

[44] E. Maskin and J. Tirole, "Markov perfect equilibrium: I. observable actions," *Journal of Economic Theory*, vol. 100, no. 2, pp. 191–219, 2001. DOI: 10.1006/jeth.2000.2785. 96

[45] J. Friedman, *Oligopoly theory*. Cambridge University Press, 1983. DOI: 10.1017/CBO9780511571893. 97

[46] Y. Narahari, D. Garg, R. Narayanam, and H. Prakash, *Game theoretic problems in network economics and mechanism design solutions*. Springer, 2009. 97, 99, 100, 101, 103

[47] R. Gibbens, R. Mason, and R. Steinberg, "Internet service classes under competition," *IEEE Journal on Selected Areas in Communications*, vol. 18, no. 12, pp. 2490–2498, 2000. DOI: 10.1109/49.898732. 108

[48] L. Duan, J. Huang, and B. Shou, "Duopoly competition in dynamic spectrum leasing and pricing," *IEEE Transactions on Mobile Computing*, vol. 11, no. 11, pp. 1706– 1719, November 2012. DOI: 10.1109/TMC.2011.213. 116

[49] R. Cornes and T. Sandler, *The theory of externalities, public goods, and club goods*. Cambridge University Press, 1996. DOI: 10.1017/CBO9781139174312. 119

[50] L. Blume and D. Easley, *The new Palgrave dictionary of economics*. Palgram Macmillan, 2008. 119

[51] A. Pigou, *The economics of welfare*. Transaction Publishers, 1924. 122, 123

[52] R. Coase, The problem of social cost. *The Journal of Law and Economics*, vol. 3, 1960. 123, 124

[53] M. Katz and C. Shapiro, "Network externalities, competition, and compatibility," *The American economic review*, vol. 75, no. 3, pp. 424–440, 1985. 120, 124, 125, 126

[54] S. Liebowitz and S. Margolis, "Network externality: An uncommon tragedy," *The Journal of Economic Perspectives*, vol. 8, no. 2, pp. 133–150, 1994. DOI: 10.1257/jep.8.2.133. 120, 124, 125, 126

[55] J. Rohlfs, "A theory of interdependent demand for a communications service," *The Bell Journal of Economics and Management Science*, pp. 16–37, 1974. DOI: 10.2307/3003090. 125

[56] B. Metcalfe, "Metcalfe's law: A network becomes more valuable as it reaches more users," *Infoworld*, vol. 17, no. 40, pp. 53–54, 1995. 126

[57] B. Briscoe, A. Odlyzko, and B. Tilly, "Metcalfe's law is wrong-communications networks increase in value as they add members-but by how much?" *IEEE Spectrum*, vol. 43, no. 7, pp. 34–39, 2006. DOI: 10.1109/MSPEC.2006.1653003. 126, 138

[58] J. Huang, R. Berry, and M. L. Honig, "Performance of distributed utility-based power control for wireless ad hoc networks," in *IEEE Military Communications Conference*, Atlantic City, NJ, USA, October 2005. DOI: 10.1109/MILCOM.2005.1606040. 131

[59] D. M. Topkis, *Supermodularity and Complementarity*. Princeton University Press, 1998. 132, 133

[60] E. Altman and Z. Altman, "S-modular games and power control in wireless networks," *IEEE Transactions on Automatic Control*, vol. 48, no. 5, pp. 839–842, May 2003. DOI: 10.1109/TAC.2003.811264. 133

[61] P. Milgrom and J. Roberts, "Rationalizability, learning and equilibrium in games with strategic complementarities," *Econometrica*, vol. 58, no. 6, pp. 1255–1277, 1990. DOI: 10.2307/2938316. 133

[62] M. Chiang, "Balancing transport and physical layers in wireless multihop networks: Jointly optimal congestion control and power control," *IEEE Journal on Selected Areas in Communications*, vol. 23, no. 1, pp. 104– 116, January 2005. DOI: 10.1109/JSAC.2004.837347. 136, 137

[63] J. Huang, R. Berry, and M. L. Honig, "Distributed interference compensation for wireless networks," *IEEE Journal on Selected Areas in Communications*, vol. 24, no. 5, pp. 1074–1084, May 2006. DOI: 10.1109/JSAC.2006.872889. 144

[64] R. Myerson, "Optimal auction design," *Mathematics of operations research*, pp. 58–73, 1981. DOI: 10.1287/moor.6.1.58. 145

[65] J. Huang, R. Berry, and M. Honig, "Auction-based spectrum sharing," *Mobile Networks and Applications*, vol. 11, no. 3, pp. 405–418, 2006. DOI: 10.1007/s11036-006-5192-y. 146

[66] J. Huang, Z. Han, M. Chiang, and H. Poor, "Auction-based resource allocation for cooperative communications," *IEEE Journal on Selected Areas in Communications*, vol. 26, no. 7, pp. 1226–1237, 2008. DOI: 10.1109/JSAC.2008.080919. 146

[67] X. Li, P. Xu, S. Tang, and X. Chu, "Spectrum bidding in wireless networks and related," *Computing and Combinatorics*, pp. 558–567, 2008. DOI: 10.1007/978-3-540-69733-6_55. 146

[68] S. Gandhi, C. Buragohain, L. Cao, H. Zheng, and S. Suri, "A general framework for wireless spectrum auctions," in *IEEE International Symposium on New Frontiers in Dynamic Spectrum Access Networks (DySPAN)*. Dublin, Ireland, 2007, DOI: 10.1109/DYSPAN.2007.12. 146

[69] X. Zhou, S. Gandhi, S. Suri, and H. Zheng, "eBay in the sky: strategy-proof wireless spectrum auctions," in *ACM MOBICOM*, Hong Kong, 2008. DOI: 10.1145/1409944.1409947. 146

[70] X. Zhou and H. Zheng, "TRUST: A general framework for truthful double spectrum auctions," in *IEEE INFOCOM*, Rio de Janeiro, Brazil, 2009. DOI: 10.1109/INFCOM.2009.5062011. 146

[71] S. Wang, P. Xu, X. Xu, S. Tang, X. Li, and X. Liu, "TODA: truthful online double auction for spectrum allocation in wireless networks," in *IEEE Symposium on New Frontiers in Dynamic Spectrum Access Networks (DySPAN)*, Singapore, 2010. DOI: 10.1109/DYSPAN.2010.5457905. 146

[72] L. Gao, Y. Xu, and X. Wang, "MAP: Multi-auctioneer progressive auction in dynamic spectrum access," *IEEE Transactions on Mobile Computing*, vol. 10, no. 8, pp. 1144–1161, 2011. DOI: 10.1109/TMC.2010.220. 146

[73] L. Gao, X. Wang, Y. Xu, and Q. Zhang, "Spectrum trading in cognitive radio networks: A contract-theoretic modeling approach," *IEEE Journal of Selected Areas in Communications*, vol. 29, no. 4, pp. 843–855, 2011. DOI: 10.1109/JSAC.2011.110415. 147

[74] D. Kalathil and R. Jain, "Spectrum sharing through contracts," in *IEEE Dynamic Spectrum Access Networks (DySPAN)*, Singapore, 2010. DOI: 10.1109/DYSPAN.2010.5457899. 147

[75] L. Duan, L. Gao, and J. Huang, "Cooperative spectrum sharing: A contract-based approach," *IEEE Transactions on Mobile Computing*, forthcoming. DOI: 10.1109/TMC.2012.231. 147

[76] G. Kasbekar, S. Sarkar, K. Kar, P. K. Muthusamy, and A. Gupta, "Dynamic contract trading in spectrum markets," in *Allerton Conference*, 2010. DOI: 10.1109/ALLERTON.2010.5706988. 147

[77] P. Muthuswamy, K. Kar, A. Gupta, S. Sarkar, and G. Kasbekar, "Portfolio optimization in secondary spectrum markets," in *International Symposium on Modeling and Optimization in Mobile, Ad Hoc, and Wireless Networks*, Princeton, USA, 2011. DOI: 10.1109/WIOPT.2011.5930023. 147

[78] L. Gao, J. Huang, Y.-J. Chen, and B. Shou, "An integrated contract and auction design for secondary spectrum trading," *IEEE Journal on Selected Areas in Communications*, vol. 31, no. 3, pp. 581–592, 2013. DOI: 10.1109/JSAC.2013.130322. 148

[79] J. Nash Jr, "The bargaining problem," *Econometrica: Journal of the Econometric Society*, pp. 155–162, 1950. 148

[80] A. Rubinstein, "Perfect equilibrium in a bargaining model," *Econometrica: Journal of the Econometric Society*, pp. 97–109, 1982. DOI: 10.2307/1912531. 148

[81] S. Chae and P. Heidhues, "A group bargaining solution," *Mathematical Social Sciences*, vol. 48, no. 1, pp. 37–53, 2004. DOI: 10.1016/j.mathsocsci.2003.11.002. 148

[82] S. Dobbelaere and R.I. Luttens, "Collective bargaining under non-binding contracts," Working paper, 2011. DOI: 10.2139/ssrn.1763693. 148

[83] S. Chae and H. Moulin, "Bargaining among groups: an axiomatic viewpoint," *Working Papers*, 2004. DOI: 10.1007/s00182-009-0157-6. 148

[84] S. Chae and P. Heidhues, "Buyers' alliances for bargaining power," *Journal of Economics & Management Strategy*, vol. 13, no. 4, pp. 731–754, 2004. DOI: 10.1111/j.1430-9134.2004.00030.x. 148

[85] S. Moresi, S. Salop, and Y. Sarafidis, "A model of ordered bargaining with applications," Working paper, 2008. 148

[86] D. Li, "One-to-many bargaining with endogenous protocol," Technical Report, The Chinese University of Hong Kong, 2010. 148

[87] Z. Zhang, J. Shi, H. Chen, M. Guizani, and P. Qiu, "A cooperation strategy based on nash bargaining solution in cooperative relay networks," *IEEE Transactions on Vehicular Technology,*, vol. 57, no. 4, pp. 2570–2577, 2008. DOI: 10.1109/TVT.2007.912960. 148

[88] L. Cao and H. Zheng, "Distributed spectrum allocation via local bargaining," in *IEEE SECON*, Santa Clara, California, 2005. 148

[89] Z. Han, Z. Ji, and K. Liu, "Fair multiuser channel allocation for ofdma networks using nash bargaining solutions and coalitions," *IEEE Transactions on Communications*, vol. 53, no. 8, pp. 1366–1376, 2005. DOI: 10.1109/TCOMM.2005.852826. 149

[90] Y. Yang, J. Huang, and J. Wang, "Dynamic bargaining for relay-based cooperative spectrum sharing," *IEEE Journal on Selected Areas in Communications*, forthcoming.
DOI: 10.1109/JSAC.2013.130812. 149

[91] H. Boche and M. Schubert, "Nash bargaining and proportional fairness for wireless systems," *IEEE/ACM Transactions on Networking*, vol. 17, no. 5, pp. 1453–1466, 2009.
DOI: 10.1109/TNET.2009.2026645. 149

# Authors' Biographies

## JIANWEI HUANG

**Jianwei Huang** is an Associate Professor in the Department of Information Engineering at the Chinese University of Hong Kong. He received a B.E. in Information Engineering from Southeast University (Nanjing, Jiangsu, China) in 2000, an M.S. and Ph.D. in Electrical and Computer Engineering from Northwestern University in 2003 and 2005, respectively. He worked as a Postdoc Research Associate in the Department of Electrical Engineering at Princeton University during 2005-2007. He was a visiting scholar at Ecole Polytechnique Federale De Lausanne (EPFL) in June 2009 and at University of California–Berkeley in August 2010. He is a Guest Professor of Nanjing University of Posts and Telecommunications.

Dr. Huang leads the Network Communications and Economics Lab (ncel.ie.cuhk.edu.hk), with the main research focus on nonlinear optimization and game theoretical analysis of communication networks, especially on network economics, cognitive radio networks, and smart grid. He is the recipient of the IEEE WiOpt Best Paper Award in 2013, the IEEE SmartGirdComm Best Paper Award in 2012, the IEEE Marconi Prize Paper Award in Wireless Communications in 2011, the International Conference on Wireless Internet Best Paper Award 2011, the IEEE GLOBECOM Best Paper Award in 2010, the IEEE ComSoc Asia-Pacific Outstanding Young Researcher Award in 2009, and Asia-Pacific Conference on Communications Best Paper Award in 2009.

Dr. Huang has served as Editor of *IEEE Journal on Selected Areas in Communications - Cognitive Radio Series*, Editor of *IEEE Transactions on Wireless Communications*, Guest Editor of *IEEE Journal on Selected Areas in Communications* special issue on "Economics of Communication Networks and Systems," Lead Guest Editor of *IEEE Journal of Selected Areas in Communications* special issue on "Game Theory in Communication Systems," Lead Guest Editor of *IEEE Communications Magazine* Feature Topic on "Communications Network Economics," and Guest Editor of several other journals including *(Wiley) Wireless Communications and Mobile Computing, Journal of Advances in Multimedia*, and *Journal of Communications*.

Dr. Huang has served as Chair and Vice Chair of IEEE MMTC (Multimedia Communications Technical Committee), the steering committee member of *IEEE Transactions on Multimedia and IEEE ICME*, the TPC Co-Chair of IEEE GLOBEBOM Selected Areas in Communications Symposium (Game Theory for Communications Track) 2013, the TPC Co-Chair of IEEE WiOpt (International Symposium on Modeling and Optimization in Mobile, Ad Hoc, and Wireless Networks) 2012, the Publicity Co-Chair of IEEE Communications Theory Workshop 2012, the TPC Co-Chair of IEEE ICCC Communication Theory and Security Symposium 2012, the Student

Activities Co-Chair of IEEE WiOpt 2011, the TPC Co-Chair of IEEE GlOBECOM Wireless Communications Symposium 2010, the TPC Co-Chair of IWCMC (the International Wireless Communications and Mobile Computing) Mobile Computing Symposium 2010, and the TPC Co-Chair of GameNets (the International Conference on Game Theory for Networks) 2009. He is also TPC member of leading conferences such as INFOCOM, MobiHoc, ICC, GLBOECOM, DySPAN, WiOpt, NetEcon, and WCNC. He is a senior member of IEEE.

# LIN GAO

**Lin Gao** is a Postdoc Research Associate in the Department of Information Engineering at the Chinese University of Hong Kong. He received the B.S. degree in Information Engineering from Nanjing University of Posts and Telecommunications in 2002, and the M.S. and Ph.D. degrees in Electronic Engineering from Shanghai Jiao Tong University in 2006 and 2010, respectively. His research interests are in the area of wireless communications and communication theory, in particular, MIMO and OFDM techniques, cooperative communications, multi-hop relay networks, cognitive radio networks, wireless resource allocation, network economics, and game theoretical models.